CHEMISTRY RESEARCH AND APPLICATIONS

MAGNETITE

SYNTHESIS, USES AND BIOLOGICAL OCCURRENCES

Chemistry Research and Applications

Additional books in this series can be found on Nova's website under the Series tab.

Additional e-books in this series can be found on Nova's website under the e-book tab.

CHEMISTRY RESEARCH AND APPLICATIONS

MAGNETITE

SYNTHESIS, USES AND BIOLOGICAL OCCURRENCES

JACQUELINE WARD
EDITOR

nova publishers
New York

For permission to use material from this book please contact us:
Telephone 631-231-7269; Fax 631-231-8175
Web Site: http://www.novapublishers.com

Library of Congress Cataloging-in-Publication Data

Magnetite (Ward)
Magnetite : synthesis, uses and biological occurrences / [edited by] Jacqueline Ward.
pages cm. -- (Chemistry research and applications)
Includes bibliographical references and index.
ISBN: 978-1-63321-591-7 (hardcover)
1. Magnetite. I. Ward, Jacqueline, 1968- II. Title.
QE391.M22M335 2014
661'.0621--dc23 2014027188

Published by Nova Science Publishers, Inc. † New York

CONTENTS

PREFACE

Magnetite is one of the two most important iron ores, the other being hematite. Although hematite is more abundant world-wide, magnetite is more widely used in North America. Some of the notable iron ranges in the U.S. include the Mesabi Range in Minnesota and the Marquette and Menominee Ranges of Michigan. Magnetite is a black, opaque, magnetic mineral, and its spinal crystal structure contains both the ferrous (Fe+2) and ferric (Fe+3) forms of iron ions, and exhibits a ferromagnetic property. Magnetite ore is upgraded by magnetic separation from a low-grade taconite ore. This book discusses the synthesis, uses and biological occurrences of magnetite.

Chapter 1 - A naturally occurring magnetite mineral called taconite is currently the most widely used raw material for producing steel in the U.S. Up to 70 % of all the steel produced in the U.S. comes from this mineral. Taconite is a low-grade iron ore with 20 – 30% iron content, from which upgraded magnetite powder, called magnetite concentrate, that contains 67-68% iron is obtained. To be used in the predominant current ironmaking process, the blast furnace process, this concentrate must first be made into green pellets with the aid of a binder followed by firing at a high temperature to produce porous but strong pellets. Many of the alternate ironmaking processes developed in recent years also use these pellets.

The magnetite concentrate is also the raw material for a transformational ironmaking technology that is being developed. This technology is based on the direct gaseous reduction of iron oxide concentrate in a flash reduction process, with the ultimate objective of significantly reducing energy consumption (by some 3.5 GJ/metric ton of iron) and environmental emissions, especially CO_2 emissions, versus the conventional blast furnace ironmaking route. The novel Flash Ironmaking Process uses gaseous reducing agents, such as natural gas, hydrogen, or syngas.

Chapter 2 - Magnetite-based solids are promising materials for a wide range of applications, including catalysis, biotechnology/biomedicine and environmental remediation. In particular, they are gaining importance as catalysts for industrial wastewater treatment. The abatement of non-biodegradable organic pollutants contained in wastewaters is a long-standing problem of environmental relevance. Fenton oxidation is a cost-effective process for the removal of those pollutants. It is based on the catalytic decomposition of hydrogen peroxide by iron salts at acidic pH, giving rise to highly reactive hydroxyl radicals. Although this technology has received major attention due to its simplicity of design, implementation and operation, it presents a significant drawback: the catalyst cannot be retained in the process, requiring further separation to prevent secondary water pollution. In this sense, the use of solid iron-containing catalysts offers a practical solution. In particular, ferromagnetic catalysts are receiving increasing attention since they can be easily separated and recovered from the liquid phase by applying an external magnetic field.

In this chapter, the authors offer a critical review of the application of magnetite-based materials to industrial wastewater decontamination by heterogeneous Fenton oxidation, also known as catalytic wet peroxide oxidation (CWPO). A survey of the catalysts investigated so far is presented focused on their nature and competitive features in terms of activity and durability. The current trends in the exploration of natural magnetic minerals, in situ-produced magnetic materials and ferromagnetic nanoparticles as new catalysts for CWPO is summarized and discussed. The author's recent results involving own-prepared ferromagnetic catalysts and natural minerals are also included.

Chapter 3 - The oldest magnet, magnetite is the prototype material of spinel iron oxides which have been applied to an antenna, magnetic core, and magnetic recording medium. Since the magnetite film has not only the nature of the magnet, but electrical conductivity at room temperature, so thus the magnetite film having sub-micron to nanometer thickness deposited on the single crystal substrate, is expected as a useful material for the next-generation magnetic devices such as magnetic random access memory (MRAM).

In this chapter, the authors research about magnetite thin films and its application are discussed. The high quality magnetite crystalline (epitaxial) films could be fabricated using a high-frequency magnetron sputtering method in the argon atmosphere with the oxygen-amount adjustment for the iron oxide target material.

Magnetite films show the very interesting physical properties; peculiar magnetic anisotropy (a nature that magnetization is easy to align along the crystal axis), and the metal-insulator transition (drastic changes in electrical resistance around 120K) depending upon their crystal properties. At first, interesting properties of the magnetite thin film are discussed. In the magnetic multilayer laminated in a few nanometers period composed of a non-magnetic and ferromagnetic material, a phenomenon whose electrical resistance changes drastically by an external magnetic field, so called "giant magnetoresistance (GMR)" effect is observed, and their multilayers are applied to the hard disk (HDD) read head having a high recording density over 100Gb/inch^2. In the magnetic multilayer using the magnetite as an electrode, the tunnel magnetoresistance (TMR) effect and large output (MR ratio over 10%) are observed at room temperature. Additionally, multilayers, called "spin valve", that are composed of magnetite as the magnetic pinning layer adjacent to the other ferromagnetic pinned layer such as iron, nickel, and cobalt, show large magnetoresistance (MR), that results in large output (several 10% MR ratio). In the second, the characteristics of magnetic multilayer using the magnetite thin film are discussed.

These multilayers using magnetite films also show very interesting magnetization processes (the external magnetic field dependence of the spontaneous magnetization), since the strong magnetic interaction between ferromagnetic metal and magnetite layer occurs in magnetoresistive multilayers, especially in spin valves. Finally, unusual magnetization reversals that could be seen in the magnetic multilayer using the magnetite thin film are also discussed.

Chapter 4 - This chapter outlines the fundamental properties of magnetite that makes it suitable for use in various biomedical applications. Magnetic properties of magnetite dispersed at nano and micro level are presented. It is well known that ferri or ferro magnetic materials with decreased particle size, at reduced nanometric range, become single domain and gain superparamagnetic behavior. In order to make use of these materials in biomedical applications, the particle agglomeration must be avoided after removing the polarizing field. This is the reason for presenting the theoretical concept of superparamagnetism. In addition, the efforts to extend the superparamagnetic behavior from nanometric level to micrometric level by a different synthesis route will be presented. Extending this property from nanometric to micrometric scale is a real challenge and it may be very useful in the biomedical field, as microparticles are expected to be safer than nanoparticles due to their lower rate of diffusion in the tissue as well as to lower reactivity and weaker interaction with the immune system. A presentation of the main advantages and disadvantages of using ultrasmall (USPIO) or small (SPIO) nano-particles and micro-particles (MPIO) in different medical applications is underlined. In this chapter, the author's own results on the synthesis of single crystalline micrometric iron oxide with superparamagnetic behavior (generically named SCMPIO) will be presented along with proposed medical uses.

The authors will focus on applications of nano/micro magnetite particles in modern biomedicine (diagnostic and therapy technologies) such as: magnetic resonance contrast agents, drug delivery, gene delivery, hyperthermia for treatment of cancer, intracellular hyperthermia under high-frequency alternating magnetic field, cellular MRI, monitoring cell migration for cell therapy, multimodal cancer therapy, immunomagnetic separation of cells, detection, immobilization and modification of biologically active compounds, cell labeling, and magnetic separation of cells and bio molecule.

Chapter 5 - A facile method of preparation of stabilized and functionalized nano- and microparticles of magnetite by successive application of oppositely charged polymers with a regular structure of macromolecular chains (chitosan and maleic acid copolymers) onto Fe_3O_4 core are developed. This approach makes it possible to create two types of magnetite interpolyelectrolyte shells, containing carboxylic or amino groups in outer layer of shell. Composition and magnetic properties of composite particles depend on nature of the copolymer of maleic acid, reaction conditions and size of obtained particles. The carboxylic groups of copolymer in outer layer of interpolyelectrolyte shell were converted into reactive anhydride groups by heating. Thermal treatment also leads to covalent cross-linking of shell and improves stability of composites both in acidic and alkaline media. Horseradish peroxidase was successfully immobilized onto covalently cross-linked and activated microparticles of Fe_3O_4 in aqueous medium without of condensing agents. The proposed reproducible and low-cost technique does not use toxic reagents or solvents at all stages, including preparation of Fe_3O_4, formation, activation and cross-linking, of magnetite shell, modification of activated surface of composite particles.

In: Magnetite
Editor: Jacqueline Ward
ISBN: 978-1-63321-591-7

Chapter 1

A NOVEL FLASH IRONMAKING TECHNOLOGY BASED ON NATURALLY OCCURRING MAGNETITE MINERAL

Hong Yong Sohn
Department of Metallurgical Engineering
University of Utah, Salt Lake City, Utah, US

ABSTRACT

A naturally occurring magnetite mineral called taconite is currently the most widely used raw material for producing steel in the U.S. Up to 70 % of all the steel produced in the U.S. comes from this mineral. Taconite is a low-grade iron ore with 20 – 30% iron content, from which upgraded magnetite powder, called magnetite concentrate, that contains 67-68% iron is obtained. To be used in the predominant current ironmaking process, the blast furnace process, this concentrate must first be made into green pellets with the aid of a binder followed by firing at a high temperature to produce porous but strong pellets. Many of the alternate ironmaking processes developed in recent years also use these pellets.

The magnetite concentrate is also the raw material for a transformational ironmaking technology that is being developed. This technology is based on the direct gaseous reduction of iron oxide concentrate in a flash reduction process, with the ultimate objective of significantly reducing energy consumption (by some 3.5 GJ/metric ton of iron) and environmental emissions, especially CO_2 emissions, versus the conventional blast furnace ironmaking route. The novel Flash Ironmaking Process uses gaseous reducing agents, such as natural gas, hydrogen, or syngas.

Keywords: Carbon dioxide, coke, concentrate, energy, flash ironmaking, hydrogen, magnetite, natural gas, reformerless, taconite

1. Basic Properties and Natural Sources of Magnetite

Magnetite is one of the two most important iron ores, the other being hematite. Although hematite is more abundant world-wide, magnetite is more widely used in North America. Some of the notable iron ranges in the U.S. include the Mesabi Range in Minnesota and the Marquette and Menominee Ranges of Michigan. Magnetite is a black, opaque, magnetic mineral, and its spinal crystal structure contains both the ferrous (Fe^{+2}) and ferric (Fe^{+3}) forms of iron ions, and exhibits a ferromagnetic property. Magnetite ore is upgraded by magnetic separation from a low-grade taconite ore.

Some of the basic properties of magnetite are listed below:

- Chemical formula: Fe_3O_4 ($Fe^{2+}Fe^{3+}{}_2O_4$) [IUPAC chemical name: iron(II,III) oxide. Common chemical name: ferrous-ferric oxide.]
- Specific Gravity: 5.18
- Mohs Hardness: 5.5 - 6.5
- Vickers Hardness: 681 - 792 kg/mm^2
- Color: gray, brown to black
- Luster: metallic to sub-metalic
- Transparency: opaque
- Magnetism: Strongly ferromagnetic.
- Curie temperature: 858 K (585 °C)

Magnetite commonly occurs in various igneous, metamorphic and sedimentary rocks, typically as disseminated crystals or grains at small weight fractions of their host rock. The discussion of natural occurrence of magnetite mineral here will be limited to the magnetite ores that are used as a raw material for iron and steelmaking. The major ore that contains commercially viable quantities of magnetite mineral in North America is taconite, which is a low-grade iron ore (25 - 30% Fe). It is a variety of chert, a fine-grained sedimentary rock containing magnetite, hematite, and silica, and occurs in the Lake Superior region.

2. Concentration and Current Ironmaking Processes Using Magnetite Mineral

Because the magnetite mineral is distributed as fine grains in large amounts of silica and other gangue minerals in taconite, a higher grade iron ore needs to be obtained from it by the process of concentration. The first step of concentration is grinding the taconite ore mixed with water into fine particles of about the size of magnetite mineral, typically 80% < 45 μm, so that the magnetite mineral may be 'liberated' (separated) from the gangue minerals. The finely ground particles are subjected to a magnetic field to separate the magnetically susceptible magnetite particles from the gangue minerals that are not affected by it. The wet magnetic portion, called the 'concentrate', contains 67-68% iron and is used for producing iron.

A typical composition of magnetite concentrate is given in Table I.

Table 1. Chemical composition of typical iron ore concentrate

Component	Wt.%
Total Fe	68
FeO	28
SiO_2	4
CaO	0.2
MgO	0.3
Al_2O_3	0.06
TiO_2	0.02
MnO	0.02
Na_2O	0.01
K_2O	0.04
P	0.01
S	0.0005

To be used in the blast furnace or a number of other alternate processes, the iron concentrate is turned into strong pellets. First, it is mixed with a binder, such as bentonite, and limestone as a flux. Then, it is rolled into green pellets 8–18 mm in diameter that contain approximately 65% iron and fired at a high temperature of 1200-1350°C. The fired pellets must remain porous enough to have high rates of reduction and yet strong enough to support the load above without being broken into small pieces. The large presence of too small pieces of solids restricts the flow of process gas through the blast furnace and also causes operating difficulties in other alternate ironmaking processes.

The blast furnace has a long history of industrial operation and has essentially reached its limit of technical improvement. The reader is referred to numerous references in the literature for further details on the blast furnace (e.g. U. S. Steel Corp., 1985; Biswas, 1981). More than 90% of iron is currently produced by the blast furnace process, while the balance is produced by direct reduction processes [IISI, 2007]. Despite the improvements in the modern blast furnace (BF) process such as increasing campaign life and productivity, decreasing coke dependence, developing improved cokemaking processes, increasing coal injection, and injecting natural gas and plastics, and despite the fact that the blast furnace is a highly energy efficient and chemically efficient reactor, it still suffers from drawbacks. The process requires the feed in the form of sinters or pellets and coke manufactured from high-grade coking coal which consumes much energy and is problematic from an environmental perspective. Also, the BF process is highly capital and energy intensive requiring large-scale infrastructure and operation. Those constraints limit the flexibility of the BF process in operation and choice of materials (Chatterjee, 1993; Manning and Fruehan, 2001).

Direct reduction processes are a group of alternate ironmaking processes in which iron oxide materials, most often iron ore pellets, are reduced in solid state at a temperature < 1,000°C. The product from direct reduction is called direct reduced iron (DRI, also called sponge iron due to its porous nature) and contains from 90 to 97 percent pure iron, the rest being mainly carbon with small amounts of other impurities. Major technologies that are currently used commercially include the MIDREX process (Atsushi et al., 2010) and the Energiron process (Duarte et al., 2010) which use a countercurrent shaft furnace to reduce iron oxide pellets, and the rotary hearth process like the ITmk3 process (Kikuchi et al., 2010). These alternate ironmaking processes currently produce almost 10% of total iron production. The SL/RN process utilizes rotary kiln to reduce lump ore and pellets with coal. This process suffers from relatively large heat loss and requires a large facility size, and hence finds limited applications.

Most of these alternate ironmaking processes, however, are not sufficiently intensive to replace the blast furnace because they cannot be operated at high temperatures due to the sticking and fusion of particles. The shaft furnace processes, which are the dominant types among these, require pelletization of concentrate with the accompanying costs and environmental problems, and also suffer from pellet disintegration problems.

3. A Novel Green Ironmaking Technology

Steel is used in a wide variety of manufactured products, such as skyscrapers, automobiles, and bridges. In order to add new steelmaking capacity to meet ever-growing demand, the traditional option of building a blast furnace and basic oxygen furnace has several drawbacks: it requires a significant capital investment, requires additional energy to make coke from special type of coal and prepare the raw iron ore as sinter and pellets, and consumes large amounts of carbon which is emitted as the greenhouse gas *carbon dioxide*. Existing alternative processes can avoid some of these issues, but are limited to low production capacities or lower quality products. Therefore, the U.S. steel industry would benefit from the development of a low capital cost process, which is scalable to large capacities, can take advantage of the availability of inexpensive iron ore concentrate, and can use fuels that significantly reduce potentially harmful greenhouse gas emissions.

Ironmaking is an energy intensive process and is one of the largest producers of greenhouse gas emissions, specifically CO_2. Unfortunately, after decades of refinement, the traditional technology is approaching its practical efficiency limits. In addition, the existing plants are in need of replacement. A typical steel mill produces in excess of 3 million tons of iron per year and thus requires immense capital investment.

An ideal process would replace the blast furnace and coke oven; would directly use iron ore concentrates for reasons presented below; and would greatly reduce energy requirements and CO_2 emissions. It should also require much less capital investment than the blast furnace/coke oven combination, and must be capable of producing 5,000 – 10,000 tons of iron per day so that it can support existing steel mills.

The steel industry contributes around 5~7 % to total anthropogenic emission of CO_2 (Tacke and Steffen, 2004). The novel Flash Ironmaking Technology, conceived by Sohn [2007], has the potential of significantly reducing energy consumption by up to 20-35% and lowering carbon dioxide emissions by 36-97% compared with the average blast furnace ironmaking route, depending on whether natural gas or hydrogen is used as the fuel/reductant. These benefits come largely from the elimination of cokemaking and sintering/pelletization, by using gaseous reduction of iron oxide concentrate in a flash reactor. It takes advantage of the large specific area of fine particles of the concentrate, much like the flash smelting of copper or nickel sulfide concentrate. The process would use inexpensive, abundant natural gas or syngas to produce hot reducing gas. It would also directly use magnetite concentrates that are currently widely used in the steel industry. The availability of large quantities of iron ore concentrate produced in the U.S. and elsewhere is also a backbone for the possible development of this technology. About 60-70% of the U.S. iron production is based on taconite concentrates (50-55 million tons/year) (USGS, 2007). Newer reserves also need to be ground down to concentrate sizes in order to beneficiate them and make them acceptable feed materials for ironmaking.

Historically, processes based on iron ore 'fines' have been less than successful. It must, however, be recognized that concentrate particles are much finer, by more than one or two orders of magnitude, than the iron ore 'fines' used in a typical fluidized bed, and the new process is meant to take advantage of this fact. The fluid bed processes use particles in the range of +0.1 mm to -10 mm (Iron Carbide, Circored and FIOR) compared with 0.02 – 0.06 mm particles in the concentrates. Thus, compared with concentrate particles, even ore 'fines' take much longer time to reduce than the time available in a flash reduction process. As will be shown below, more than 95% reduction of iron oxide concentrate particles could be obtained by hydrogen or a mixture of carbon monoxide and hydrogen in several seconds at temperatures of 1300°C or above.

The flash ironmaking process would reduce energy consumption up to 20% compared with blast furnace by using iron oxide concentrates that do not require pelletization or sintering, as required in other ironmaking processes (Sohn, 2007; USGS, 2007; Sohn and Choi, 2010), and by eliminating the use of coke ovens. Greenhouse gas emissions would be significantly reduced by using natural gas or hydrogen as the reducing agent instead of coke. Estimates show that the use of natural gas would emit 39% less carbon dioxide than with a blast furnace-based process (Pinegar et al., 2012), while the use of pure hydrogen would emit almost no carbon dioxide in the ironmaking step, depending on the method of its production.

When natural gas is used as the reducing agent, carbon can be added into the molten iron product, lowering its melting point and facilitating direct steelmaking in the same unit to maximize energy savings. Alternatively, the product could also be collected as a solid to be transported to an Electric Arc Furnace for steelmaking. Future system designs could add a correct amount of carbon for added benefits such as inexpensive energy generation, foamy slags that reduce heat loss, and improved tap-to-tap time.

Another major advantage of flash ironmaking over existing ironmaking processes that use shaft or fluidized-bed furnaces is the elimination of sticking and particle fusion at high temperatures. The ability to use ore fines provides a cost advantage over processes that require ore to be consolidated into pellets for ironmaking. The fine particles also cut the furnace's processing time to mere seconds. Other potential benefits include improved refractory life, and ease of feeding raw materials into the reactor. Thus, the flash ironmaking process could provide steel plants with significantly more energy-efficient and customized iron production facilities and could eventually replace blast furnace operations as we know them today and other alternate ironmaking processes.

Below, we will discuss general description of the novel ironmaking process and the results of the kinetic feasibility test which forms a major justification of the development, followed by a larger scale laboratory test of the process. Work on flow sheet development, process simulation, and economic feasibility analysis will then be presented together with a description of on-going even larger scale test work.

3.1. Flash Ironmaking

Development of sustainable metallurgical processes and the fusion of ideas from different disciplines for technology innovation are two most important factors in modernizing the metallurgical industry. This section describes the development of a novel flash ironmaking process. In this process, natural gas or hydrogen is partially combusted with industrial oxygen through a burner on top of the flash reactor to generate a reducing gas stream at 1600 – 1900 K. Iron ore concentrate is injected in the vicinity of the burner and undergoes reduction reaction as it flows down. In one version of the process the reduced product is collected as solid particles and briquetted to be used in separate steelmaking furnaces. In another version, the reduced iron is collected as a molten bath as part of a continuous direct steelmaking process, as sketched in Figure 1.

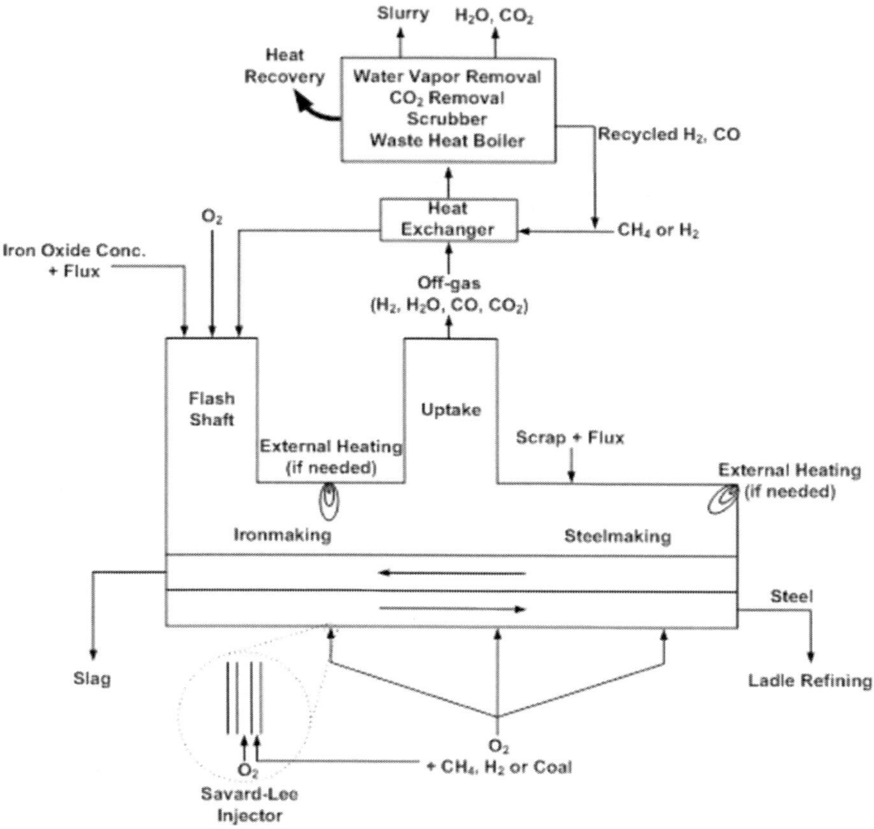

Figure 1. A schematic diagram of the novel flash ironmaking process. {Adapted from Mohassab-Ahmed and Sohn [2013]].

The Flash Ironmaking Technology is initially expected to be operated using natural gas as a fuel and reducing agent. Natural gas is an abundant readily available resource in the U.S. and elsewhere in the world. According to the 2011 Annual Energy Outlook (U.S. Energy Information Administration, 2011), the estimations on the U.S. production rate of natural gas predict a steady increase. In addition, there is a big push to increase natural gas production in the U.S. from its considerable reserves. The recently discovered reserves in Marcellus Shale (Pittsburgh Post-Gazette, 2011) represent a good example of this trend.

For an annual production of 20 million tons of iron, which would represent about 40% of the current U.S. production rate, the Flash Ironmaking Process would require 0.33 TCF/year (trillion cubic feet per year) of natural gas. This represents less than 1.5% of the total U.S. consumption rate (22.7 TCF/year) (U.S. Energy Information Administration, 2011). As a comparison, this is similar to the 1.5% of the total U.S. energy consumption used by the U.S. steel industry. When natural gas is used as a reducing agent, the proposed process will produce iron with varying carbon contents such as the iron produced by HYL's ZR self-reforming process (Duarte et al., 2010). In the latter process, carbon levels can be up to 5.5 %.

A critical condition for the proposed process is whether iron oxide concentrate can be reduced to a high metallization degree within the few seconds of residence time available in a typical flash furnace. This issue will be addressed below. Suffice it to note here that the rate is sufficiently fast for the reduction of iron ore concentrate in a flash reaction process above 1450 K (Sohn et al., 2009; Wang and Sohn, 2013).

3.2. Reduction Kinetics of Magnetite Concentrate Particles

Hydrogen is the main reducing agent in the flash ironmaking process that takes place at temperatures above 1473 K, even when partial combustion of natural gas producing an H_2+CO mixture is used. Most previous work was done on the reduction of iron ore particles at temperatures below 1373 K, which is much lower than the temperature at which the flash ironmaking process is expected to operate (Sohn, 2007; Sohn et al., 2009; Pinegar et al., 2012). What is more, none of the rate equations reported in the literature contained all the information needed for reactor design relevant to the flash ironmaking process. Thus, a comprehensive kinetics expression applicable to the flash ironmaking conditions was necessary.

Such reduction kinetics of the gaseous reduction of magnetite concentrates was measured and a rate equation that contains the effects of process conditions in a flash ironmaking reactor was developed by Wang and Sohn [2013]. These conditions include temperature, partial pressures of hydrogen and water vapor, and particle size. Furthermore, data covering the entire fractional reduction degree vs. time, which had not been available for the conditions applicable to the flash ironmaking process, were generated in their work. An additional emphasis on the formulation was that such a rate expression be readily applicable to the development of an overall process model for the design and simulation of a reactor for the flash ironmaking technology.

A. Materials

Magnetite concentrate from a taconite ore of the Mesabi Range was used in the study by Wang and Sohn [2013]. The concentrate particles were irregularly shaped and angular, as shown in Figure 2. The chemical composition for each size fraction is presented in Table II.

Table II. Chemical composition (wt%) for each size fraction of magnetite concentrate

Component	Particle Size		
	20-25 μm	32-38 μm	45-53 μm
Total Fe	71.0	70.3	67.6
FeO	30.6	30.2	29.7
P	0.01	0.01	0.01
S	0.02	0.02	0.02
C	0.25	0.28	0.58
Sr	0.02	0.01	0.02
SiO_2	1.68	2.08	4.51
Al_2O_3	0.13	0.10	0.17
CaO	0.21	0.35	0.85
MgO	0.11	0.18	0.49
MnO	0.09	0.12	0.24
Cr_2O_3	0.10	0.07	0.11
K_2O	0.01	0.01	0.01
Na_2O	0.10	0.10	0.10
TiO_2	0.01	0.01	0.01
ZrO_2	0.01	0.03	0.03
CuO	N/A	N/A	N/A

Most of the iron oxide was magnetite and the total iron content ranged from 68 to 71%. The concentrate was screened to 25 - 32, 32 - 38, and 45 - 53 μm fractions for kinetics measurements in a drop-tube reactor.

Figure 2. SEM micrographs - Upper Row: SEM micrographs of screened particles; Lower Row: SEM micrographs of polished sections. {Adapted from Wang and Sohn [2013].}

B. Experimental Apparatus and Procedure

The reduction rate of fine concentrate particles by hydrogen above 1473 K took only a few seconds to be completed (Sohn et al., 2009). It is extremely difficult to record weight change of the sample in such a short reduction time and avoid diffusion resistance among particles in the sample layer in a crucible by using a customary TGA system. Thus, a high temperature drop-tube reactor system was used (Choi and Sohn, 2010) for accurate determination of the reduction rate of individual concentrate particles. It utilizes a dilute particles-gas conveyed system to measure the chemical reaction rate of fine particles entrained in a reducing gas.

As shown in Figure 3, this apparatus consisted of a vertical high temperature drop-tube furnace, a pneumatic powder feeder, gas delivery lines, a powder cooling and collecting system, and an off-gas outlet. The furnace system used a vertical alumina tube (5.6 cm ID, 193 cm long), in which a reaction zone was maintained at a constant temperature of 1173 to 1773 K by bar-type SiC elements. A cylindrical alumina honey-comb was inserted in the tube and hung right above the beginning of the reaction zone as a flow straightener and a heat exchanger for the reducing gas.

The pneumatic powder feeding system consisted of a syringe pump, a vibrator, a carrier gas line, a powder container, and a powder delivery line. Dried screened concentrate was charged in a powder container, a disposable test tube, which was held by a bore-through Swagelok and sealed with an O-ring. The mixture of hydrogen and nitrogen was fed into the test tube (0.9 cm ID) and passed through the powder delivery line (0.12 cm ID) at the top of the powder feeder continuously entraining a small amount of the powder. The feeding tube was vibrated by an electric vibrator to prevent clogging. The vial was pushed up by a motor at a constant target advancing rate which was determined based on the calibration data between the advancing rate of the syringe pump and the powder feed rate. The reacted powder was collected in a powder collector at the bottom of the reactor.

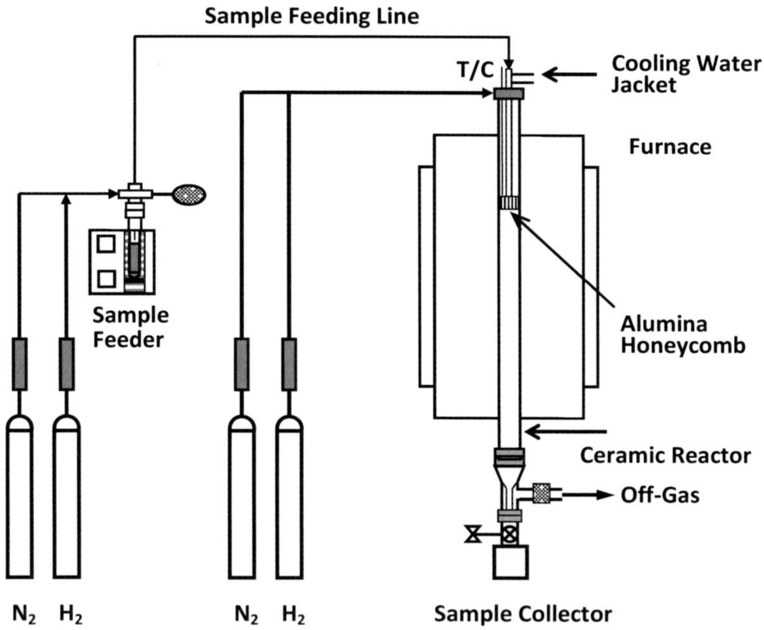

Figure 3. Schematic diagram of drop-tube reactor system. {Adapted from Wang and Sohn [2013].}

At a designed experimental temperature, hydrogen and nitrogen were charged into the reactor at predetermined flow rates after the furnace system was purged by nitrogen. The concentrate powder was charged into the reactor after the temperature inside the furnace became stable.

C. Sample Analysis

The reduction degree from raw concentrate toward metallic iron (Fe) was determined by chemically analyzing the product. The total iron content was determined by titration based on international standard on the determination of metallic iron in direct reduced iron (DRI) (ISO, 2006). To confirm the analytical procedure, either hematite or iron powder with over 99% purity was also analyzed along with the samples. For further characterization, a Siemens D5000 X-ray diffractometer was used for the compositional analysis of the samples. A TOPCON SM-300 Scanning Electron Microscope (SEM) equipped with an energy dispersive spectrometer (EDS) was used to examine the microstructure of the concentrate and the quantitative elemental analysis. An Optical Microscope was used to observe the polished sections of the samples.

D. Definition of Parameters

To accurately determine the kinetics, it is best to carry out the experiment under a condition in which the gaseous reactant concentration remains constant over the entire reactor length. This requires a sufficiently large excess of the gaseous reactant over the stoichiometric amount. The term % excess H_2 used here is defined as follows, especially taking into consideration the fact that the final stage of the iron oxide reduction, i.e. the reaction of FeO with H_2 or CO, is significantly limited by equilibrium. For example, at 1400°C,

$$FeO(s) + H_2(g) = Fe(s) + H_2O(g) \qquad \Delta G^o \approx +900\,cal \quad (1400^o C) \tag{1a}$$

$$[PH_2/(PH_2 + PH_2O)]_{eq} = 0.5 \tag{1b}$$

This reaction has a slightly positive standard Gibbs free energy, and the equilibrium gas product has an H_2/H_2O molar ratio of ~ 1 at this temperature, i.e. 50% H_2 and 50% H_2O when pure hydrogen is used as the reductant. Taking the equilibrium composition into consideration, % excess H_2 was defined as follows (Choi and Sohn: 2010; Wang and Sohn, 2013):

$$Fe_3O_4 + H_2 = 3FeO + H_2O \tag{2a}$$

$$FeO + H_2 = Fe + H_2O \tag{2b}$$

Overall: $Fe_3O_4 + 4H_2 = 3Fe + 4H_2O$ (2c)

$$K_{R2} = \left[\frac{p_{H_2O}}{p_{H_2}}\right]_{eq} = \left[\frac{n_{H_2O}}{n_{H_2}}\right]_{eq} \tag{3}$$

$$n_{H_2,\min} = n_o^i + \frac{n_o^i}{K_{R2}} = \left[1 + \frac{1}{K_{R2}}\right] n_o^i \tag{4}$$

$$\%\ excess\ \ H_2 = \frac{n_{H_2,supplied} - n_{H_2,\min}}{n_{H_2,\min}} \times 100 \tag{5}$$

Reaction (2a), the hydrogen reduction of magnetite to wustite, has a large equilibrium constant, i.e. essentially irreversible, whereas Reaction (2b) is considerably limited by chemical equilibrium. Thus, the amount of hydrogen in equilibrium with the water vapour present in the gas was used for the equilibrium constant of Reaction (2b) represented by Equation (3). The minimum amount of hydrogen, $n_{H2,min}$, then is the amount of hydrogen used to remove the oxygen from the iron oxide, n^i_o, plus the hydrogen required by equilibrium to be present with the water vapour produced by the reduction reaction, n^i_o / K_{R2}. The % excess H_2 was then calculated from the total amount of hydrogen fed into the reactor compared with the minimum amount of hydrogen, $n_{H2,min}$, as indicated in Equation (5).

The fractional reduction is obtained from the difference of the mass of oxygen combined with iron in the concentrate particles before and after reduction. The percent reduction is calculated as follows (Wang and Sohn, 2013).

$$\%\text{Reduction} = \frac{m_{o(0)} - m_{o(t)}}{m_{o(0)}} \times 100\% \tag{6}$$

$$m_{o(0)} = \frac{m_{TFe(t)}}{(\%TFe)_0} \times (\%O)_0 \tag{7}$$

$$m_{o(t)} = m - m_{TFe(t)} - \frac{m_{TFe(t)}}{(\%TFe)_t} \times [100 - (\%TFe)_0 - (\%O)_0] \tag{8}$$

Here, m is the mass of reduced sample used in the titration, $m_{o(t)}$ and $m_{o(0)}$ are, respectively, the mass of the removable oxygen in the reduced sample collected after reaction for time t, and the corresponding mass of the removable oxygen in the unreduced dry concentrate. $m_{TFe(t)}$ is the mass of total iron in the reduced sample (mass is m), which is determined by titration. $(\%TFe)_0$ and $(\%O)_0$ are the mass percentage of total Fe and removable oxygen of the corresponding unreduced concentrate, which are determined respectively by titration and complete reduction conducted in a separate horizontal tubular furnace. The weight change related to possible volatile species such as phosphorus and sulphur was neglected based on the small contents as given in Table II and by the negligible weight loss confirmed by heating under an inert atmosphere.

E. Determination of Residence Time

The value of residence time was calculated from the length of the hot zone, which started from the tip of the powder feeding tube, the linear velocity of the gas, and the terminal falling velocity for the creeping flow region expressed by the Stokes' law assuming that particles fall at a constant velocity in the isothermal zone. As the gas flows downward after the flow straightener, a change in the flow mode from a plug flow to a fully developed laminar flow is expected in the very short length of the circular tube. From the normalized development length relationship suggested by Durst et al. [2005] as indicated in Equation (9), it was found that the fully developed state of the flow is reached in less than 5% of the hot zone length.

$$L/D = [(0.619)^{1.6} + (0.0567Re)^{1.6}]^{1/1.6} \tag{9}$$

where L = length of the isothermal zone, D = diameter of tubular reactor, Re = Reynolds number.

Since the solid particles fall along the centerline, the residence time in this work was calculated by taking the maximum velocity, which is twice the average velocity (= volumetric flow rate divided by cross-sectional area), as the linear velocity of gas. The relevant equations are:

$$u_t = d_p^2 g(\rho_p - \rho_g)/18\mu \tag{10}$$

$$u_p = u_g + u_t \tag{11}$$

$$t = L/u_p \tag{12}$$

where, all in consistent units, d_p= particle size, g = gravitational acceleration, ρ_p = particle density, ρ_g = gas density, μ = viscosity of gas, u_p = particle velocity relative to tube wall, u_g = centerline gas velocity at furnace temperature, u_t = terminal velocity of a falling particle and t = residence time.

F. Rate Equation

It was demonstrated by detailed calculation (Choi, 2010) that the reduction rate controlled by external-mass-transfer + pore-diffusion is much greater than the measured rates. Therefore, the overall reaction was rate-controlled by the reaction of individual particles and not affected by pore diffusion or external mass transfer.

The nucleation and growth kinetics expression at a constant temperature given by the following equation was found to best describe the reduction rate of magnetite concentrate particles:

$$[-Ln\ (1-X)]^{1/2} = \left[k \cdot f_p\left(p_{H_2}, p_{H_2O}\right) \cdot f_d\left(d_p\right)\right] \cdot t = k_{app} \cdot t \tag{13}$$

where X = fractional reduction degree, t = reaction time, k = rate constant, d_p = particle size, $f_p\left(p_{H_2}, p_{H_2O}\right)$ = concentration driving force, $f_d\ (d_p)$ = rate dependence on particle size, and k_{app} = apparent rate constant.

The reduction of magnetite to iron proceeds through the formation of FeO. However, for small particles, it is extremely difficult to follow the irregular evolution of morphology involving FeO formation and the overlapping regions of different reduction extents. Furthermore, the reduction of Fe_3O_4 to FeO accounts only for the first 25 % of conversion, whereas the main interest in this work and the flash ironmaking technology is the attainment of high degrees of reduction. Thus, Equation (13) was developed as a global rate equation for the overall reduction process.

The following expression for pressure dependence, which satisfies the equilibrium condition as a special case when the net rate is zero, was found applicable with K being the equilibrium constant of the reaction ($FeO + H_2 = Fe + H_2O$):

$$f_p\left(p_{H_2}, p_{H_2O}\right) = p_{H_2}^{1/2} - \left(p_{H_2O}/K\right)^{1/2} \tag{14}$$

In this reaction system, the reduction rate increased with the particle size at 1423 and 1473 K within the size range tested. The effect was negligible above 1573 K. It is believed that the major factor in this size dependence is the variation in the contents of gangue materials (Wang and Sohn, 2013). The contents of the main gangue materials CaO, MgO and SiO_2 in iron oxide concentrate particles were higher for larger particles. Previous investigations have shown that the addition of CaO (Iguchi and Inouye, 1982; Shigematsu and Iwai, 1989) or MgO (McCune and Wynblatt, 1983; Tasker et al., 1985) increases the rates of gaseous reduction of wustite. What is more, cracks tend to be formed in the large particles during the reduction more readily than in the smaller particles, and the dimension of solid between cracks can be smaller than the size of the smaller particles. The rate dependence on the particle size was assumed to take the form of $(d_p)^s$. Thus,

$$f_d(d_p) = (d_p)^s \tag{15}$$

with

$$s = \frac{2.73\times10^8}{T^2} - \frac{345930}{T} + 109.5;\ T = 1423\sim1573\ \text{K} \tag{16a}$$

$$s = 0;\ T \geq 1573 \tag{16b}$$

After the functional dependence of the rate on partial pressures of gaseous species (f_p) and particle size (f_d) were determined, the reaction rate constant k was obtained as follows:

$$k = (9.9 \times 10^{14}) \cdot exp(-\frac{463000}{RT}) \tag{17}$$

resulting in the following conversion-vs.-time relationship:

$$[-Ln\ (1 - X)]^{1/2} = (9.9 \times 10^{14}) \cdot e^{-\frac{463000}{RT}} \cdot \left[p_{H_2}^{1/2} - (p_{H_2O}/K)^{1/2}\right]_{am} \cdot (d_p)^s \cdot t \tag{18}$$

where R is 8.314 J/mol/K, T is in K, p is in atm, d_p is in μm, and t is in seconds.

The instantaneous rate of reaction, which is valid under varying temperature and gas composition, can now be expressed as

$$\frac{dX}{dt} = 2\,k \cdot \left[p_{H_2}^{1/2} - (p_{H_2O}/K)^{1/2}\right] \cdot \left(d_p\right)^s \cdot (1-X) \cdot [-Ln\,(1-X)]^{1/2} \tag{19}$$

This equation satisfactorily represents the experimentally determined rate of magnetite concentrate, as a sample set of data given in Figure 4 indicates (Wang and Sohn, 2013).

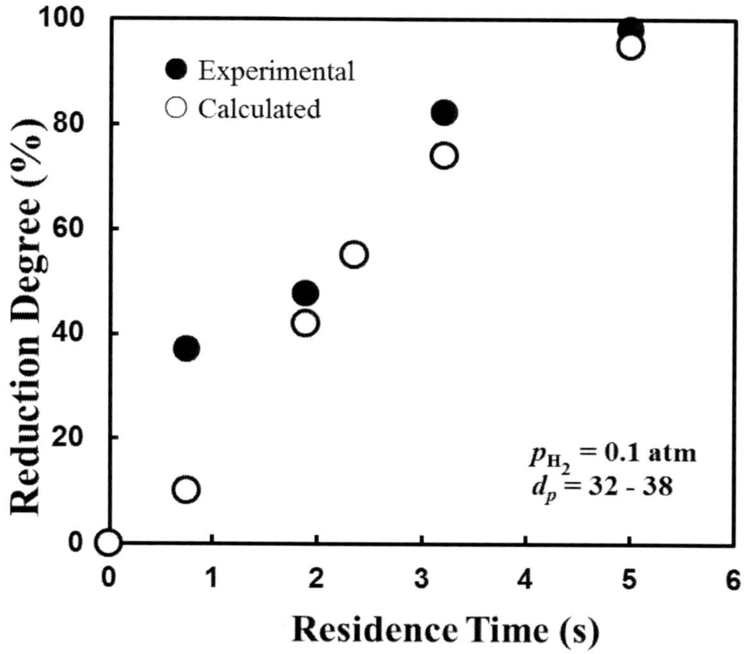

Figure 4. Comparison of experimental and calculated results. p_{H2} = 0.1 atm; d_p = 32 - 38 μm; 1400°C. {Adapted from Wang and Sohn [2013].}

The rate measurements indicated that magnetite concentrate particles can be reduced to a sufficiently high degree of metallization within 2~7 seconds of residence time available in a flash reactor.

G. Reduction by H_2 + CO

The kinetics of magnetite concentrate reduction with H_2+CO gas mixtures, which would be generated by partial combustion of natural gas, was also performed (Sohn et al., 2014a). Figure 5 presents the effect of temperature on the reduction of a magnetite concentrate with particle size of 20-25 μm by H_2+CO+N_2 gas mixtures at a residence time of 5 seconds. It can be seen that more than 95 % reduction was achieved at 1300 °C in 5 seconds with a 25%H_2-25%CO-50%N_2 gas mixture. This result may suggest that CO contributes only moderately to the reduction of magnetite concentrate particles at 1300 °C.

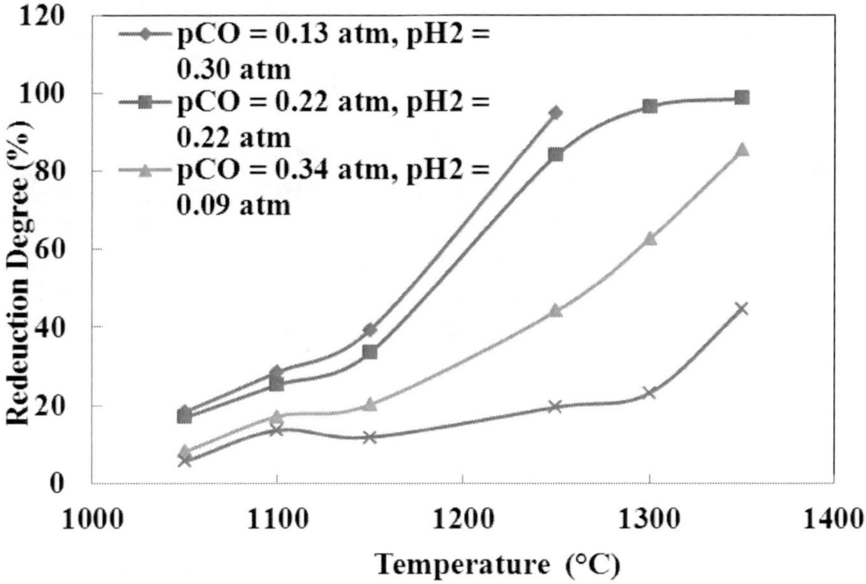

Figure 5. The effect of temperature on the reduction of magnetite concentrate (20-25 μm) by H_2+CO+N_2 gas mixtures at a residence time of 5 seconds. {Adapted from Sohn et al. [2014a].}

3.3. Tests in a Laboratory Flash Reactor

Based on the above results, larger scale suspension reduction tests were conducted by Sohn et al. [2014b] in a laboratory flash reactor shown in Figure 6.

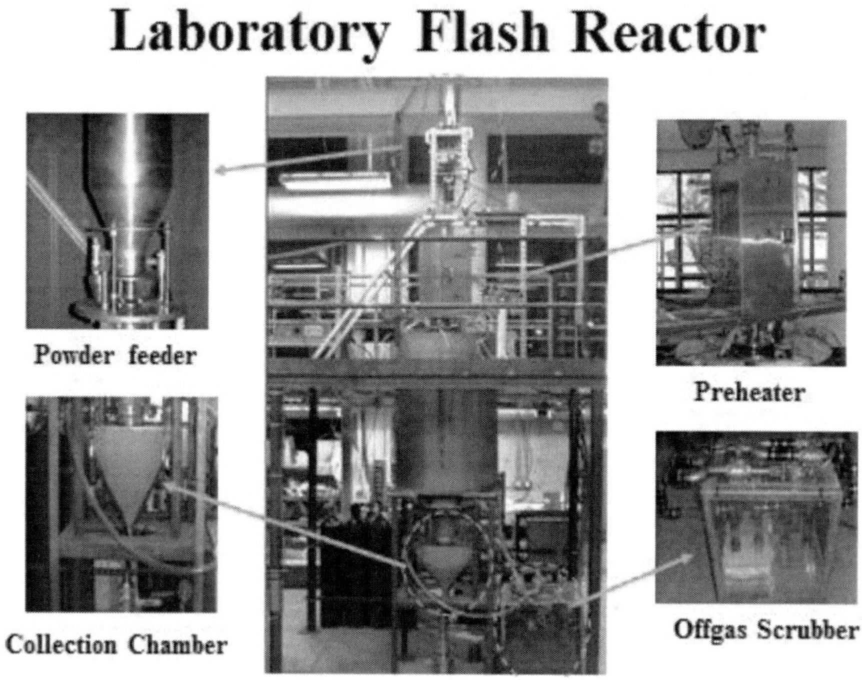

Figure 6. Utah laboratory flash reactor.

The apparatus consists of five subsystems; a vertical furnace system, an electric power control system, a gas delivery system, a preheating system, and a powder feeding system. A tubular steel reactor (20.3 cm ID, 244 cm long) was electrically heated by six SiC heating elements, which were grouped into two and managed by two SCR controllers respectively. The maximum temperature obtained with the set-up was 1473 K in 76 cm of isothermal zone. The iron oxide concentrate particles were fed into the reactor by a pneumatic powder feeding system at about 1.5 g/min of feeding rate. The duration of hydrogen reduction of the particles was determined by the residence time of particles in the isotheral zone.

Figures 7 and 8 summarize the results of the runs using hydrogen as the fuel/reductant and feeding the concentrate particles through injection ports on the reactor roof. Figure 7 shows reduction degree versus residence time curves for fixed excess driving force whereas Figure 8 presents reduction degree versus excess driving force curves for fixed residence time.

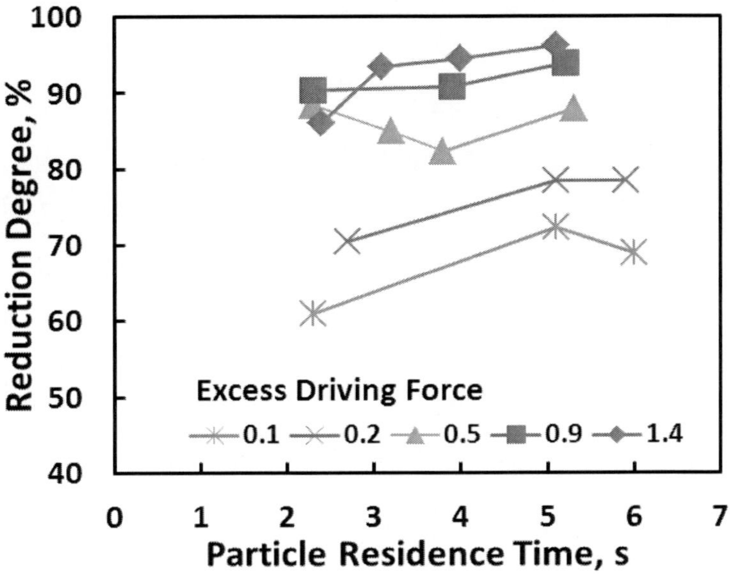

Figure 7. Reduction degree as a function of particle residence time for the powder produced in the laboratory flash reactor using hydrogen as the fuel/reductant and at temperatures of 1150-1200 °C. {Adapted from Sohn et al. [2014b].}

From Figure 7, it is noted that reduction degrees of 90 % or higher were achieved for residence times longer than 3 seconds and excess driving forces of 0.9 or higher at temperatures of 1150–1200 °C. As shown in Figure 8, reduction degrees close to 90 % were achieved with excess driving forces as low as 0.5 when the residence time was 5 seconds or longer in the same temperature range. This is a very encouraging result in view of the fact that this temperature range is considerably lower than the >1300 °C to be used in an industrial reactor.

For excess driving forces lower than 0.5, it is noted (Figure 8) that reduction degrees around 80 % were achieved. A residence time of 5 seconds yielded reduction degrees of 86 % and 78 % for excess driving forces of 0.3 and 0.2, respectively.

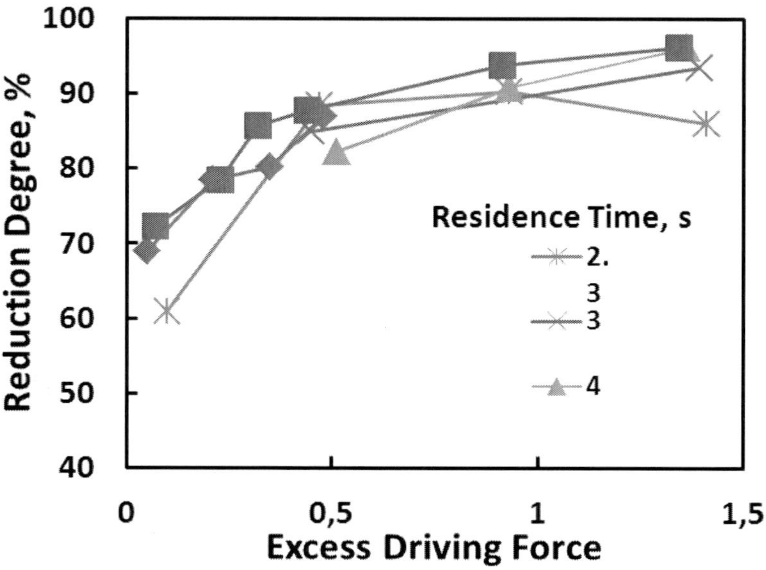

Figure 8. Reduction degree as a function of excess driving force for the powder produced in the laboratory flash reactor using hydrogen as the fuel/reductant and at temperatures of 1150-1200 °C. {Adapted from Sohn et al. [2014b].}

SEM micrographs were obtained to examine the variation of microstructure of iron oxide concentrate particles during the process of reduction. Figure 9 shows the microstructural changes with the degree of reduction increasing from 0% to 92%. From Figure 9-(a), it is seen that the raw concentrate particles are irregular both in shape and size. While the degree of reduction increases from 0% to 29%, cracks began to appear on the surface of particle as shown in Figure 9-(b). As the degree of reduction increased to 43% and 67%, more cracks and pores were formed. When the degree of reduction reached over 80%, the whole particle became porous. The porosity allows hydrogen to penetrate to the interior of the particles.

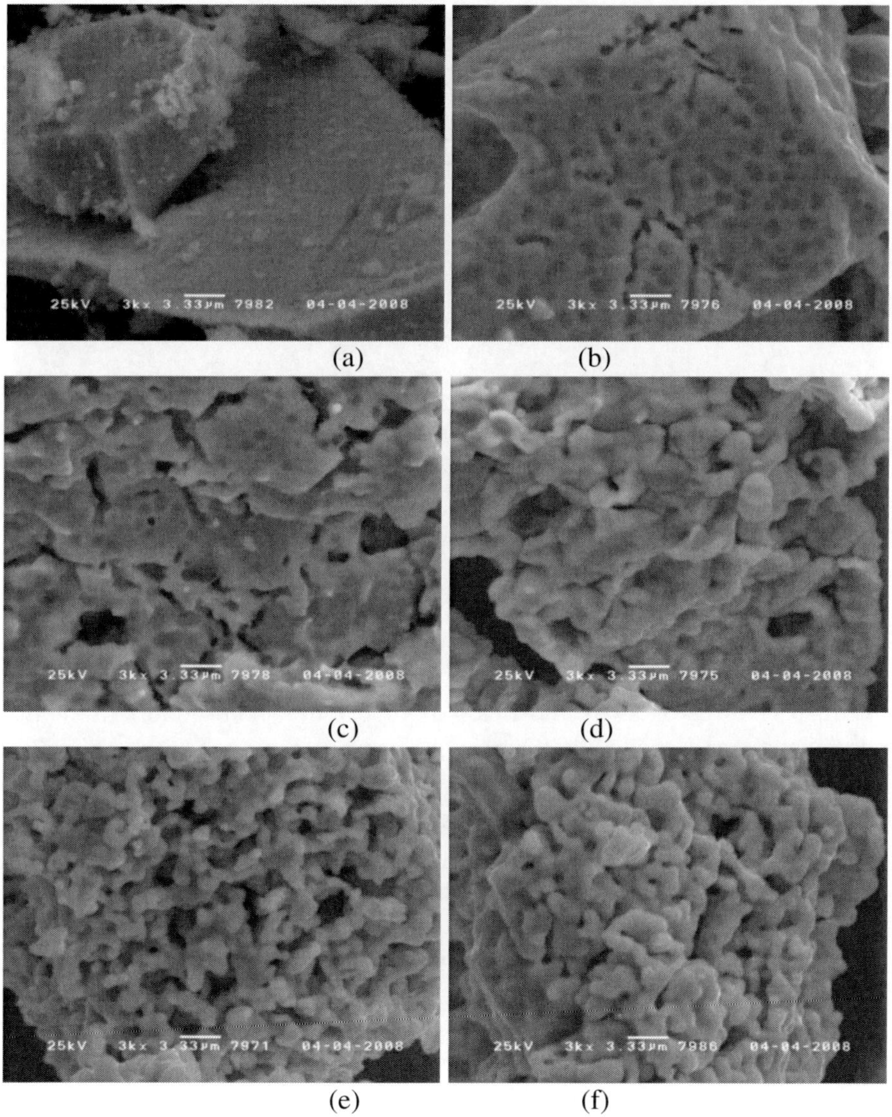

Figure 9. SEM micrograph of sample with (a) 0% reduction (raw iron oxide concentrate) (b) 29% reduction (c) 43% reduction (d) 67% reduction (e) 80% reduction and (f) 92% reduction. (1423 K) {Adapted from Sohn et al. [2009].}

However, it should be noted that the morphological change does not occur at the same rate. In Figure 10-(a), the two particles with similar size and shape show different porosity. Sometimes even smaller particle did not become porous while the larger particle became completely porous as seen in Figure 10-(b). This is because the carrier gas entraining particles was introduced as an expanding jet and thus each particle had different trajectory inside the reactor which had a certain temperature distribution, which caused the particles to experience different temperatures and residence times.

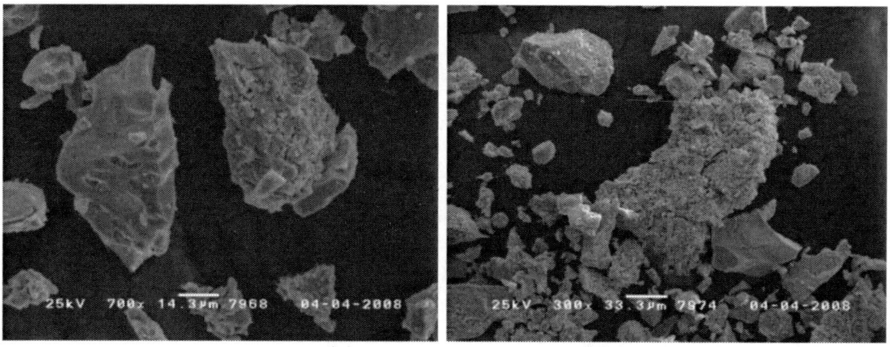

(a) (b)

Figure 10. SEM micrograph of sample with (a) 29% reduction and (b) 80% reduction (1423 K) {Adapted from Sohn et al. [2009].}

Although hydrogen is best as a reductant and/or fuel from the environmental and reduction kinetics viewpoint, it is currently expensive. Instead, the syngas which is mainly composed of H_2 and CO from the reforming of natural gas or the coal gasification has been used as a reducing gas mixture for the majority of direct reduction processes. Thus, the use of syngas as a reductant was examined in the laboratory flash reactor. The syngas was simulated with a mixture of H_2, CO, and N_2 while keeping the compositional ratio the same as that obtained by mixing hydrogen with a combustion gas from an internal burner. The input amount of each component was calculated based on material and energy balances considering the preheated gas temperature, the heat of reaction between CH_4 and O_2 in the burner, and the heat absorbed by the room-temperature hydrogen added to improve the reduction rate.

A series of experiments showed that about 90% reduction was accomplished in 3.5 seconds at 860 % excess hydrogen (32.7 kPa H_2, 53.4 kPa N_2). It should be noted that the average barometric pressure of Salt Lake City is 86.1 kPa. The % reduction remained similar when 11.6 kPa of N_2 was replaced with the same amount of CO. When 10% (7.6 kPa) of H_2 was replaced with CO the conversion decreased from 90% to 80% even at about

4.5 seconds of residence time. Although there was a decrease in reduction rate by introducing syngas instead of H_2, it is still of interest especially at higher operating temperatures.

3.4. Process Design and Simulation

Detailed flow sheets for different versions of the flash ironmaking process depending of the fuel type were developed by Pinegar et al. (2011; 2012; 2013a) by using commercially available software METSIM. Although hydrogen is the cleanest fuel/reductant for the flash ironmaking process, it is still expensive to produce. Thus, the results for using natural gas in a "reformerless" mode are summarized here (Pinegar et al., 2012).

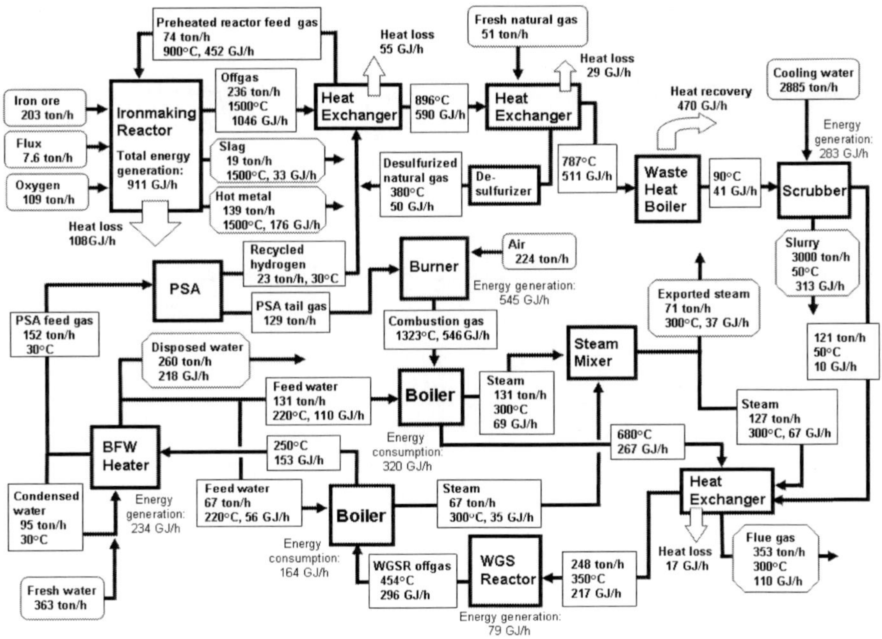

Figure 11. Material and energy flows for a reformerless ironmaking process producing 1 million tons of iron annually (300 days operation in 1 year). {Adapted from Pinegar et al. [2012].}

The reformerless process involves an ironmaking step and off-gas treatment/hydrogen recycling steps. Water gas shift (WGS) reaction and pressure swing adsorption (PSA) were incorporated in the off-gas treatment steps. A block diagram of the reformerless process is shown in Figure 11. Descriptions of unit operations and streams as well as other details are given elsewhere (Pinegar et al., 2012). The annual iron production rate was set to be 1 million tons with 300 days of annual operation. Major results of the process simulation can be summarized as follows:

- CO_2 emissions from the flash ironmaking process: 1.0 ton/ton Fe vs. 1.6 ton/ton Fe from an average blast furnace.
- Energy requirement for the flash ironmaking process: 14.7 GJ/ton Fe vs. 18.0 GJ/ton Fe for an average blast furnace. See Table III for further details, including the case of using hydrogen. (Note: These numbers are those calculated based on an approach used in DOE reports. For a discussion on different items included in 'energy requirements' for ironmaking, the reader is referred to Sohn and Martinez-Olivas [2014].)

Table III. Energy requirement for commercial-scale Flash Ironmaking Process using natural gas or hydrogen compared with that for the average blast furnace process

		Reformerless Natural Gas	Hydrogen	Blast Furnace[a]
INPUT[b] (GJ/mtFe)	Fuel combustion[c]	19.22	14.05	13.6
	Pelletizing/ Sintering/ Cokemaking			3.01/0.65 /2.02
	Limestone calcination	0.26	0.26	
	Total	19.48	14.31	19.28
OUTPUT[d] (GJ/mtFe)	Heat recovery	4.77	2.80	1.32
	Total	4.77	2.80	1.32
NET = INPUT – OUTPUT (GJ/mtFe)		14.7	11.5	18.0

[a] Calculated based on material balance following the same procedure as that for flash ironmaking.

[b] All raw materials are considered to enter the system at room temperature (25°C).

[c] Fuel combustion energy input in this case is the heat of combustion of the entire fuel + reductant.

[d] Energy output is the useful heat recovered (plus any heating values which in this case are zero because we considered fully oxidized species) in both flash ironmaking processes and blast furnace.

Economic analysis based on the simulation results indicated that the flash ironmaking process has favorable economic feasibility when operated with natural gas, and the capital investment would be significantly lower than for a flash ironmaking plant based on blast furnace and coke oven. Any carbon penalty would make the flash ironmaking technology even more attractive (Pinegar et al., 2013b).

3.5. Further Development Work

The development work is continuing at the University of Utah toward the design and construction of a large-scale bench reactor, as illustrated in Figure 12, and beyond that, a 50,000-100,000 ton of iron per year industrial pilot testing (Sohn et al., 2012). The large-scale reactor test program aims to overcome several technical difficulties before an industrially viable technology can be finalized, such as the method of supplying energy required to maintain the necessary temperature, the installation of proper refractory and insulation linings, and the safe control of the entire operation. The large-scale bench reactor will produce iron directly from iron ore concentrate (10~100 μm) using natural gas or hydrogen as the fuel as well as the reductant in a closed cylindrical reactor. Of the specific design and functional features necessary for the bench facility, the reactor heating system, which has been the major limitation in the preliminary scale-up tests, is a key issue.

The bench reactor under construction consists of a refractory-lined vertical vessel (80 cm ID, 150 cm long) in which iron oxide concentrate is to be reduced in flight by hot reducing gases produced by partial combustion of natural gas or hydrogen. The reducing gas is to be introduced through a swirl burner installed at the top part of the reaction chamber. The concentrate powder will be fed using a pneumatic feeding system through the burner and auxiliary ports located on the vessel roof. A typical feed rate of iron ore concentrate in this scale will be 5 kg/h. Concentrate particles are to be supplied through the burner or in the vicinity of the hydrogen /oxygen flame. The average temperature of the reaction shaft will be varied in the range of 1650~1850 K. The energy for the sensible heat of solid and gas feed materials and for the reduction reaction of iron oxide is to be supplied by a single burner (1.27 GJ/h) in combination with a 100 kW plasma torch as a supplementary heat source. The entire system will be monitored by several safety measures and controlled by proper emergency shut-off procedures.

Figure 12. Preliminary large-scale bench reactor facility layout. [Image courtesy of Berry Metal Company].

Figure 12 shows the conceptual diagram and preliminary facility layout of the bench-scale test facility. This system to be installed at the University of Utah has many features that resemble an eventual industrial reactor. They include a burner as the main heat source and a refractory-lined vessel. Upon completing this project phase, the engineering data necessary for industrial pilot-plant operations will be determined, such as heat loss, product collection, raw materials feeding, process control strategy, burner design, and off-gas treatment. In addition, a risk mitigation plan addressing the critical technical hurdles to be overcome has been devised, as follows:

1. Heat supply. If the necessary temperature cannot be attained because of the large heat loss typically present in a small test reactor, plasma can be used to augment the heating. Calculations indicated that the heat loss in the bench reactor could reach more than 30% of total heat input to the reactor. Similar calculations for an industrial-scale reactor (15 m ID, 15~45 m height) showed that the heat loss would drop to ~2% of total input.
2. Particle residence time. The burner is designed to provide different levels of swirl flow which allows the control of the particle residence time.

3. Refractory lining. Different types of refractory liners may be tested in the course of the experiments.

Safety will be the most important issue in this bench testing. A great deal of attention is paid to safety-related details such as the Sequence of Operation and the Emergency Shut-Off Procedures for a safe and sound operation as well as a Hazards and Operability (HAZOP) study.

CONCLUSION

Among the many occurrences and uses of magnetite, by far the largest amount is used for producing steel. For this purpose, a low grade ore taconite that contains finely dispersed grains of magnetite is used. When taconite is ground and subjected to a magnetic field, fine particles of magnetite concentrate particles are separated from the gangue minerals. Currently, this concentrate is made into pellets before being used for producing iron. The pelletization process involves fusing the fine particles together into pellets that have sufficient strengths, thereby losing a large portion of the specific surface area and thus the reactivity.

A transformational ironmaking technology under development has been described here, which directly uses the magnetite concentrate, bypassing the pelletization and sintering steps. It therefore eliminates the associated energy consumption and pollutant emissions. This technology is also based on gaseous reduction of iron oxide, utilizing natural gas or hydrogen and thus significantly reducing CO_2 emissions. Recent developments in natural gas production by the hydraulic fracturing technique greatly increases the attractiveness of flash ironmaking based on the magnetite mineral.

REFERENCES

Atsushi, M., Uemura, H., and Sakaguchi, T., *Kobelco Technology Review* 2010, Dec. no. 29, pp. 50-57. http://www.kobelco.co.jp/english/ktr/pdf/ktr_29/050-057.pdf

Biswas, A. K.; *Principles of Blast Furnace Ironmaking: theory and practice*; Cootha Pub. House; Brisbane, Australia, 1981, 528 pp.

Chatterjee, A.; *Beyond the Blast Furnace*, pp. 1, CRC press, Boca Raton, FL, 1993.

Choi, M. E.: *Suspension Hydrogen Reduction of Iron Ore Concentrate*, Ph. D. Dissertation, University of Utah, Salt Lake City, Utah, 2010.

Choi, M. E. and Sohn, H. Y. *Ironmaking Steelmaking* 2010, vol. 37(2), pp. 81-88.

Duarte, P. E.; Becerra, J.; Lizcano, C.; Martinis, A. *Steel Times International* 2010, vol. 34 (3), pp. 25 -26, 28, 30.

Durst, F.; Ray, S.; Unsal, B.; Bayoumi, O. A. *Trans. ASME* 2005, vol. 127, pp. 1154–1160.

Iguchi, Y.; Inouye, M., *Trans. ISIJ* 1982, vol. 22, pp. 679-687.

IISI, Steel Statistical Yearbook; 2007;. http://www.worldsteel.org/pictures/publicationfiles/SSY2007.pdf.

ISO - International Organization for Standardization: *Direct reduced iron-Determination of metallic iron-Bromine-methanol titrimetric method*, ISO 5416, 3rd ed., Geneva, Switzerland, 2006.

Kikuchi, S.; Ito, S.; Kobayashi, I.; Tsuge, O.; Tokuda, K. *Kobelco Technology Review* 2010, No. 29, pp. 77-84. http://www.kobelco.co.jp/english/ktr/pdf/ktr_29/077-084.pdf

Manning, C. P.; Fruehan, R. J. *JOM*, 2001, vol. 53 (10), pp. 36-43.

McCune, R. C.; Wynblatt, P. *J. American Ceramic Soc.*1983, vol. 66, pp. 111-117.

Mohassab-Ahmed, M. Y.; Sohn, H. Y. *JOM*, 2013, vol. 65 (11), pp. 1559-1565.

Pinegar, H. K.; Moats, M. S.; Sohn, H. Y. *Steel Res. Int'l.*, 2011, vol. 82 (8), pp. 951–963.

Pinegar, H. K.; Moats, M. S.; Sohn, H. Y. *Ironmaking Steelmaking*, 2012, vol. 39, pp. 398-408.

Pinegar, H. K.; Moats, M. S.; Sohn, H. Y. *Ironmaking Steelmaking*, 2013a, vol. 40, pp. 32-43.

Pinegar, H. K.; Moats, M. S.; Sohn, H. Y. *Ironmaking Steelmaking*, 2013b, vol. 40, pp. 44-49.

Pittsburgh Post-Gazette: *Penn State forecasts boom for Marcellus Shale*, 2011, http://www.post-gazette.com/pg/11202/1161933-503.stm ; see also http://www.eia.gov/analysis/studies/usshalegas/

Shigematsu, N.; Iwai, H. *ISIJ, International*, 1989, vol. 29(6), 486-494.

Sohn, H. Y. *Steel Times International*, 2007, vol. 31 (3), 68-72.

Sohn, H. Y.; Choi, M. E. *Collected Proceedings: Supplemental Proceedings: Vol. 1 Materials Processing and Properties, International Symposium on*

High-Temperature Metallurgical Processing, 139th TMS Annual Meeting, Seattle, WA, ed. J. Drelich et al. (Warrendale, PA: TMS, 2010), 347-354.

Sohn, H. Y.; Olivas-Martinez, M. in *Energy Technology 2014 Carbon Dioxide Management and Other Technologies*, 143rd Annual TMS Meeting, San Diego, California, 2014, ed.by C. Wang et al., TMS/Wiley, pp. 49-56, 2014.

Sohn, H. Y.; Choi, M. E.; Zhang, Y.; Ramos, J. E. *Iron Steel Technology (AISI Trans.)* 2009, vol. 6 (8), pp. 158-165.

Sohn, H. Y.; Choi, M. E.; Olivas-Martinez, M. *Proceedings of 6th Intl Congress on Sci. and Tech. of Ironmaking (ICSTI '12)*, Oct. 14 – 18, 2012, Rio de Janeiro, Brazil, ISSN 2176-3135, Paper No. 22117 (2012), 460-468.

Sohn, H. Y.; Chen, F.; Pal, J.; Mohassab-Ahmed, M. Y.; Olivas-Martinez, M. unpublished work, 2014a.

Sohn, H. Y.; Olivas-Martinez, M.; Elzohiery, M. H.; Fan, D.; Mohassab-Ahmed, M. Y. unpublished work, 2014b.

Tacke, K.-H.; Steffen, R. *Stahl Eisen*. 2004, vol. 124 (4), 45-52.

Tasker, P. W.; Colburn, E. A.; Mackrodt, W. C. *J. Amer. Ceram. Soc.*, 1985, vol. 68, 74-80.

United States Steel Corporation: *The Making, Shaping and Treating of Steel*, 10th edition, Association of Iron and Steel Engineers, 1985.

U.S. Energy Information Administration, *Annual Energy Outlook 2011*, http://www.eia.gov/forecasts/aeo/index.cfm

USGS, 2005 Minerals Yearbook - Iron Ore (Report U.S. Department of the Interior, U.S. Geological Survey, 2007).

Wang, H.; Sohn, H. Y. *Metall. Mater. Trans. B*, 2013, vol. 44B, 133-145.

In: Magnetite
Editor: Jacqueline Ward

ISBN: 978-1-63321-591-7

Chapter 2

PREPARATION AND APPLICATION OF MAGNETITE-BASED CATALYSTS TO WASTEWATER TREATMENT

Macarena Munoz*[*], *Zahara M. de Pedro, Jose A. Casas and Juan J. Rodriguez
Chemical Engineering Section, Universidad Autónoma de Madrid, Campus de Cantoblanco, Madrid, Spain

ABSTRACT

Magnetite-based solids are promising materials for a wide range of applications, including catalysis, biotechnology/biomedicine and environmental remediation. In particular, they are gaining importance as catalysts for industrial wastewater treatment. The abatement of non-biodegradable organic pollutants contained in wastewaters is a long-standing problem of environmental relevance. Fenton oxidation is a cost-effective process for the removal of those pollutants. It is based on the catalytic decomposition of hydrogen peroxide by iron salts at acidic pH, giving rise to highly reactive hydroxyl radicals. Although this technology has received major attention due to its simplicity of design, implementation and operation, it presents a significant drawback: the catalyst cannot be retained in the process, requiring further separation to

[*] Corresponding author: E-mail address: macarena.munnoz@uam.es.

prevent secondary water pollution. In this sense, the use of solid iron-containing catalysts offers a practical solution. In particular, ferromagnetic catalysts are receiving increasing attention since they can be easily separated and recovered from the liquid phase by applying an external magnetic field.

In this chapter, the authors offer a critical review of the application of magnetite-based materials to industrial wastewater decontamination by heterogeneous Fenton oxidation, also known as catalytic wet peroxide oxidation (CWPO). A survey of the catalysts investigated so far is presented focused on their nature and competitive features in terms of activity and durability. The current trends in the exploration of natural magnetic minerals, in situ-produced magnetic materials and ferromagnetic nanoparticles as new catalysts for CWPO is summarized and discussed. Our recent results involving own-prepared ferromagnetic catalysts and natural minerals are also included.

FUNDAMENTS

The Fenton process is a well-established technology for the abatement of non-biodegradable organic pollutants. It is based on the catalytic decomposition of H_2O_2 by means of iron at acid pH, giving rise to hydroxyl radicals, one of the most powerful oxidants known ($E^o = 2.73$ V). The history of Fenton chemistry dates to 1894, when Henry J. Fenton reported that H_2O_2 could be activated by Fe^{2+} salts to oxidize tartaric acid [1]. However, it was not applied as an oxidation process for the removal of organic pollutants until the late 1960s [2]. Research on the Fenton chemistry to wastewater treatment began in academic laboratories only around 1990 [3]. Since then, the number of works related to the Fenton process has increased exponentially over the years (Figure 1).

The application of the Fenton process to wastewater treatment is attractive due to the fact that hydrogen peroxide is safe and easy to handle, and poses no lasting environmental threat, since it readily decomposes to water and oxygen. Likewise, iron is inexpensive, safe and environmentally friendly. Among the advantages of the Fenton process relative to other oxidation techniques are the simplicity of equipment and the mild operation conditions; mainly for these reasons this process has been regarded as one of the most cost-efficient technologies among the Advanced Oxidation Processes [4-8]. So far, the Fenton process has been successfully applied for the treatment of highly polluted wastewaters such as cosmetics [9, 10], olive-mill [11, 12], chemicals [10], pulp and paper [6], power plant [13] or sawmills [14], among others [15].

ure 2 illustrates a typical flow scheme of the Fenton process. In a first the raw wastewater is homogenized by mechanical agitation and the pH usted by the addition of acid. Then, the wastewater is fed to the reactor, re the catalyst (iron solution) and hydrogen peroxide are directly added ce the desired temperature is reached. The optimum pH is 3, whereas the talyst and hydrogen peroxide doses as well as the temperature and the eaction time must be established upon experimentation. In general, the catalyst concentration is around 1 part iron per 4-25 parts of hydrogen peroxide (w/w) [19] while the optimum dose of oxidant is frequently the stoichiometric amount for complete mineralization [20]. Although most of the works dealing with Fenton oxidation have been carried out at ambient temperature, recent studies have shown that increasing the temperature, up to 120 °C, clearly improves both the oxidation rate and the degree of mineralization of the pollutants [21]. After the Fenton treatment the wastewater is neutralized with NaOH, and thus iron complexes precipitates as $Fe(OH)_3$. A final filtration unit to remove fine aggregates may be necessary to meet discharge requirements.

The sludge generated during the process is mechanically dewatered to as high dry solids content as possible. The cost associated to sludge treatment and disposal can represent up to 10-50% of the total operating cost of the wastewater treatment [10, 17]. Therefore, reducing the amount of sludge produced is imperative. This objective of sludge minimization has stressed the importance of using iron-bearing solid catalysts in the process. Moreover, the application of supported iron catalysts would also avoid the continuous loss of catalyst associated to the homogeneous Fenton process.

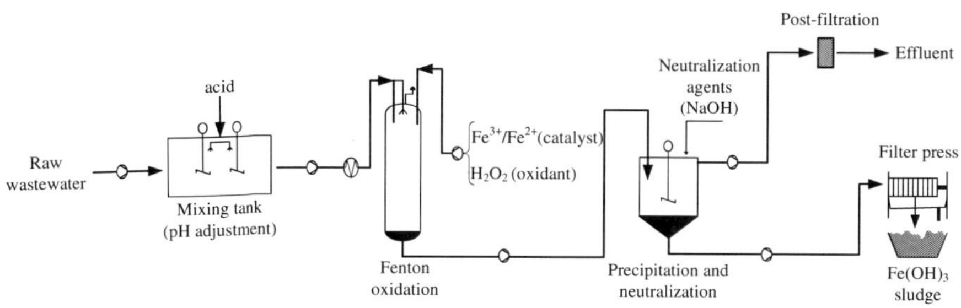

Figure 2. Schematic diagram for Fenton treatment.

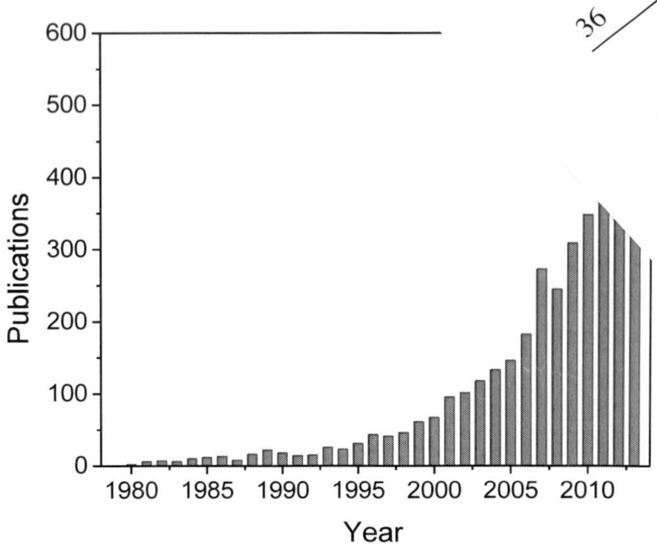

Figure 1. Evolution of the number of publications dealing with the application of Fenton oxidation to wastewater treatment. Source: *Scopus (April 2014)*.

The Fenton mechanism, which has been well established in the literature [3, 16-18], is quite complex and includes more than 20 reactions. Briefly, H_2O_2 decomposes catalytically by means of Fe^{2+} at acid pH giving rise to hydroxyl radicals. Fe^{3+} can react, at acid pH, with H_2O_2 yielding hydroperoxyl radicals, regenerating the catalyst and thus sustaining the process:

$$Fe^{2+} + H_2O_2 \rightarrow Fe^{3+} + \cdot OH + OH^- \qquad (1)$$

$$Fe^{3+} + H_2O_2 \rightarrow Fe^{2+} + \cdot OOH + H^+ \qquad (2)$$

The oxidation of organic pollutants gives rise to intermediate species, which can further react with ·OH leading to mineralization, namely complete oxidation up to CO_2, H_2O and (if the pollutant contains heteroatoms) inorganic salts. Thus, the overall process can be described by the following reactions:

$$COD + H_2O_2 \xrightarrow{Fe^{2+}/\,Fe^{3+}} reaction\ intermediates \qquad (3)$$

$$reaction\ intermediates + H_2O_2 \xrightarrow{Fe^{2+}/\,Fe^{3+}} CO_2 + H_2O + inorganic\ salts \qquad (4)$$

CATALYTIC WET PEROXIDE OXIDATION (CWPO)

Despite the advantages of homogeneous Fenton oxidation, it suffers also of some drawbacks that limit a more extended application: sludge generation, the need for pH adjustment before and after reaction and the loss of the catalyst in the effluent which cannot be recovered and requires further removal. To solve or minimize these drawbacks, the use of heterogeneous catalysts in the so-called Catalytic Wet Peroxide Oxidation (CWPO) or heterogeneous Fenton oxidation is a promising alternative. As can be seen in Figure 3, the interest of the scientific community in this subject has increased greatly in recent years.

The support plays an important role in the catalytic system. It increases the surface area of the metal or metal oxide by providing a matrix that enables their dispersion as very small particles. It inhibits the sintering of the active phase of the catalyst and improves its thermal and chemical stability. Typical supports for CWPO catalysts are presented in Figure 4a. As observed, pillared clays represent the most frequent, followed by zeolites and silica. However, low attention has been paid so far to the use of iron oxides. On the other hand, iron is by far the metal most frequently used as active phase although many others have been also tested, as can be seen in Figure 4b.

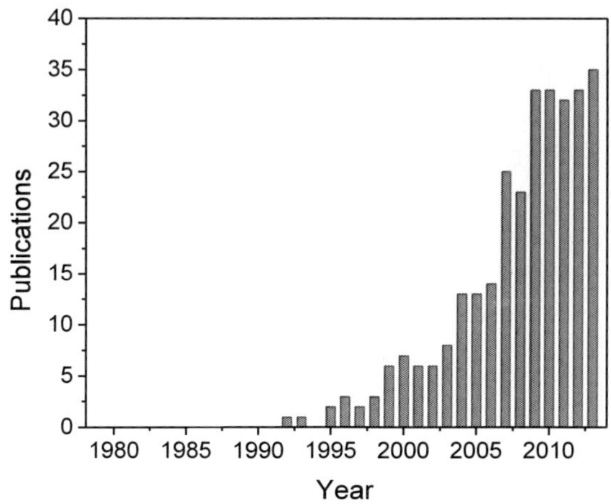

Figure 3. Evolution of the number of publications dealing with the application of CWPO to wastewater treatment. Source: *Scopus (April 2014)*.

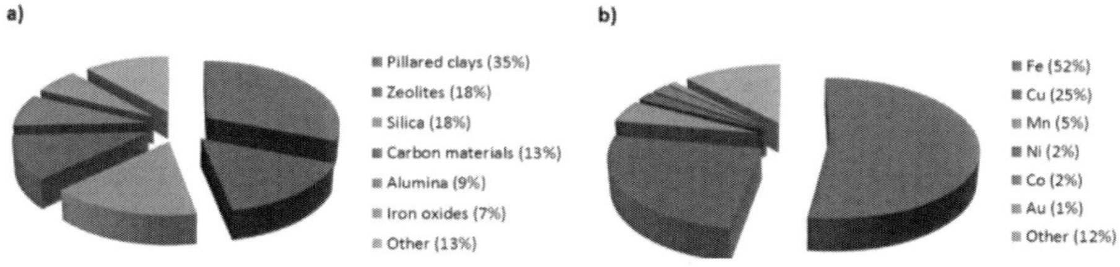

Figure 4. Literature generated in the CWPO field distributed by sort of catalytic supports (a) and active phase (b) studied. Source: *Scopus (April 2014)*.

Although CWPO represents an interesting alternative to conventional Fenton oxidation, unfortunately most of the catalysts studied so far have shown relatively low activity and stability due to the poisoning of the catalytic agents, the reduction of the catalyst surface and, overall, the leaching of iron [22-25]. These shortcomings have prevented the commercial application of CWPO to wastewater treatment. In this sense, developing stable catalysts represents one of the most important challenges in CWPO. Recently, iron supported on γ-Al_2O_3 (Fe/γ-Al_2O_3) has proved to be a highly active and stable catalyst for CWPO of phenolic compounds [26, 27] and real industrial wastewater [28]. The catalyst showed a remarkable stability in long-term continuous experiments (100 h) with limited Fe leaching (<3% of the initial loading).

The recovery and reusability of the catalyst represents a key issue regarding its potential application. Since CWPO is typically conducted in a suspension of powdered particles, an additional separation step is required to recover the catalyst from the treated water. This poses a major drawback for the application of CWPO to wastewater treatment. In this context, the development of magnetic catalysts is a promising solution since the catalyst could be easily separated upon the application of a magnetic field, thereby simplifying its recovery and reusability.

Magnetite-Based Catalysts in CWPO

Magnetite (Fe_3O_4) has attracted considerable research interests in materials chemistry because of its unique properties such as magnetic, electric, catalytic, biocompatibility and low toxicity. In particular, the use of magnetite as active phase in CWPO is receiving increasing attention due to its magnetically easy separation from the reaction system and its relatively high abundance and low cost. Moreover, it contains Fe(II) which should have a significant positive effect in the activity of the catalyst since the rate constant of Equation (1) is much higher than Equation (2) (70 $M^{-1}\cdot s^{-1}$ *vs.* 0.001 – 0.01 $M^{-1}\cdot s^{-1}$) [17]. In fact, the octahedral structure of magnetite can easily accommodate both Fe(II) and Fe(III), which means that Fe(II) can be reversibly oxidized and reduced back in the same position. Thus, the magnetite-based catalysts present a higher potential for degradation of recalcitrant pollutants than the conventional iron-supported catalysts where the active phase very majorly contains Fe(III) usually in the form of hematite (Fe_2O_3). Despite these positive characteristics scarce work has been carried out with magnetite in heterogeneous Fenton systems until a few years ago. Although several attempts studying the potential application of magnetite and other natural minerals as catalysts in Fenton oxidation were carried out previously [29, 30], the first work published devoted uniquely to magnetite as CWPO catalyst is that reported by Zhang et al. (2008)[31]: "Decomposing phenol by the hidden talent of ferromagnetic nanoparticles". These authors claimed that magnetic nanoparticles represent a promising alternative to the conventional catalysts used in CWPO due to their higher catalytic activity as well as their easily magnetic recovery and further reusability. Since then, the number of studies dealing with the use of magnetite-based catalyst in CWPO has increased noticeably (Figure 5). In general, three different trends can be distinguished regarding the application of magnetite-based materials to CWPO: natural minerals, *in situ* synthesis of magnetite from Fe-supported catalysts and magnetic nanoparticles (MNPs). Table 1 summarizes the work done regarding the application of magnetite-based catalysts so far.

Magnetic Minerals As Novel Fenton Catalysts

The use of natural minerals as catalysts in CWPO represents an interesting alternative due to their availability and low cost. The potential of iron-oxide minerals to catalyze the oxidation of organic pollutants has been scarcely studied so far although their efficiency has been clearly demonstrated [29, 30, 57, 58]. The most widely studied minerals are goethite [29, 57, 59], hematite [30, 57], ferrihydrite [30, 57], pyrite [30], lepidocrocite [30] and magnetite

[30,32]. Among them, magnetite is gaining considerable attention due to the presence of Fe(II) in its structure and its magnetic properties. Kwan and Voelker (2003)[29] studied the rates of hydroxyl radicals generation by different mineral-catalyzed Fenton systems concluding that Fe(III) oxides are catalytically less active than their Fe(II) counterparts. In this sense, Matta et al. (2007)[30] found that Fe(III) oxides (hematite, goethite, lepidocrocite and ferrihydrite) were less effective than Fe(II) minerals, such as magnetite and pyrite, in the degradation of 2,4,6-trinitrotoluene. Similarly, Hanna et al. (2008)[60] compared the catalytic activity of maghemite, magnetite and goethite for methyl orange degradation, concluding that magnetite exhibited the highest rate constants normalized to surface area per unit mass of oxide.

Due to its excellent properties, magnetite (Fe_3O_4) is by far the natural magnetic mineral most widely used as catalyst in CWPO [61]. Magnetite is a spinel iron oxide with chemical formula $(Fe(III))_{tet}[Fe(II)Fe(III)]_{oct}O_4$ where Fe(III) ions occupy equally both octahedral and tetrahedral sites and Fe(II) ions are placed only in octahedral sites. Their redox properties provide high activity in oxidation processes. So far, it has been used for the oxidation of a wide variety of organic compounds such as trinitrophenol [30], pentachlorophenol [32], phenol [62], tetrabromobisphenol-A [34], various dyes [33] and polycyclic aromatic hydrocarbons [35].

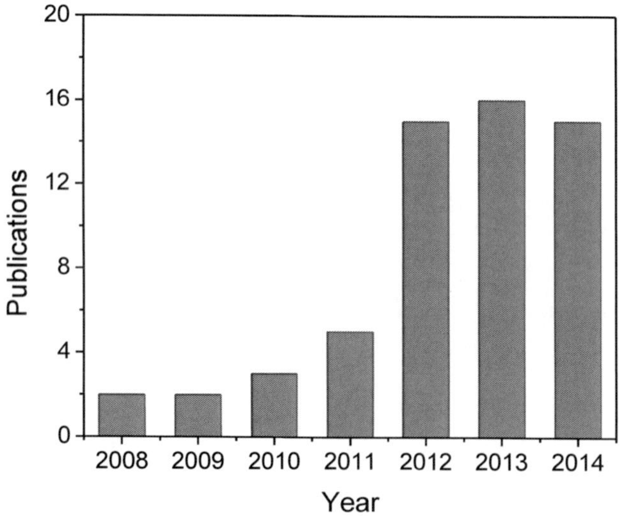

Source: Scopus (April 2014).

Figure 5. Evolution of the number of publications dealing with the application of magnetic catalysts in the Fenton process.

Table 1. Magnetite-based catalysts in CWPO for removal of non-biodegradable organic pollutants

Catalyst	Pollutant	Operating conditions	Results	Stability	Reference
MAGNETIC MINERALS					
Magnetite	Pentachloro-phenol	$C_{cont\text{-}cat\text{-}H2O2} = 0.05 - 2 - 5\ g{\cdot}L^{-1}$ T = 20 °C P = 1 atm pH = 7	$X_{cont} = 90\%$ (t = 9 h) The kinetic rate constant increased with the addition of chelating agents: EDTA>>Oxalate>CMCD>without catalyst	Iron leaching < 14 $mg{\cdot}L^{-1}$	[32]
Natural vanadium-titanium magnetite	Acid Orange II	$C_{cont\text{-}cat\text{-}H2O2} = 0.07 - 1 - 0.34\ g{\cdot}L^{-1}$ T = 25 °C P = 1 atm pH = 3	$X_{cont} = 98\%$ (t = 4 h)	3 runs $X_{cont(1st)} = 98\%$ $X_{cont(2nd)} = 93\%$ $X_{cont(3rd)} = 90\%$\	[33]
Titanomagnetite (UV/Fenton)	Tetrabromobis-phenol A	$C_{cont\text{-}cat\text{-}H2O2} = 0.02 - 0.125 - 0.34\ g{\cdot}L^{-1}$ T = 25 °C P = 1 atm pH = 6.5 (UV 6W λ=365 nm)	$X_{cont} = 100\%$ (t = 4 h)	3 runs $X_{cont(1st)} = 100\%$ $X_{cont(3rd)} = 90\%$ (t = 4 h)	[34]
Magnetite rich sandy soil (MRS)	Oil hydrocarbon	$C_{cont\text{-}cat\text{-}H2O2} = 0.4 - 100 - 10\ g{\cdot}L^{-1}$ T = 25 °C P = 1 atm pH = 6.7	$X_{cont} > 80\%$ (one week)	--	[35]

Table 1. (Continued)

Catalyst	Pollutant	Operating conditions	Results	Stability	Reference
IN SITU GROWTH OF MAGNETITE					
Magnetic Fe_2MO_4 activated carbon	Methyl orange	$C_{cont\text{-}cat\text{-}H2O2} = 0.05 – 2.5 – 0.6\ g{\cdot}L^{-1}$ T = 30 °C P = 1 atm pH = 4	$X_{cont} = 100\%$ $X_{TOC} = 59\%$ (t = 2 h)	3 runs $X_{cont(1st)} = 100\%$ $X_{cont(2nd)} = 95\%$ $X_{cont(3rd)} = 85\%$ Iron leaching: 0.47, 0.30 and 0.26 $mg{\cdot}L^{-1}$ for 1, 2 and 3 runs	[36]
Magnetite/mesocellular carbon foam	Phenol	$C_{cont\text{-}cat\text{-}H2O2} = 0.01 – 0.1 – 0.34\ g{\cdot}L^{-1}$ T = 20 °C P = 1 atm pH = 3	$X_{cont} = 95\%$ (t = 4 h)	Multiple runs Non appreciable loss of activity Iron leaching $< 0.6\ mg{\cdot}L^{-1}$	[37]
Sludge derived magnetic porous carbon	1-diazo-2-naphtol-4-sulfonic acid	$C_{cont\text{-}cat\text{-}H2O2} = 0.25 – 0.5 – 0.5\ g{\cdot}L^{-1}$ T = 25 °C P = 1 atm pH = 5	$X_{cont} = 97\%$ $X_{TOC} = 87\%$ (t = 4 h)	3 runs $X_{cont(1st)} = 97\%$ $X_{cont(1st)} = 94\%$ $X_{cont(1st)} = 90\%$	[38]
Ferromagnetic $\gamma\text{-}Al_2O_3$-supported iron catalyst ($Fe_3O_4/\gamma\text{-}Al_2O_3$)	4-chlorophenol (4-CP) 2,4-dichlorophenol (2,4-DCP) 2,4,6-trichlorophenol (2,4,6-TCP)	$C_{cont\text{-}cat\text{-}H2O2} = 0.1 – 1.0 – 0.35$ (4-CP), 0.21 (2,4-DCP), 0.18 (2,4,6-TCP) $g{\cdot}L^{-1}$ T = 50 °C P = 1 atm pH = 3	$X_{cont} = 100\%$ (1, 2, 3 h for 4-CP, 2,4-DCP and 2,4,6-TCP, respectively) $X_{TOC} > 75\%$ (t = 4h)	Long-term continuous experiment (100 h). No loss of activity. Iron leaching $<2.5\ mg{\cdot}L^{-1}$	[27]

Catalyst	Pollutant	Operating conditions	Results	Stability	Reference
Reactive Fe particles dispersed in a carbon matrix	Methylene blue	$C_{cont\text{-}cat\text{-}H2O2} = 0.2 – 4.3 – 86\ g \cdot L^{-1}$ T = 26 ºC P = 1 atm pH = 6	$X_{cont} = 95\%$ (t = 3 h)	--	[39]
MAGNETIC NANOPARTICLES					
Ferromagnetic nanoparticles (MNPs)	Phenol	$C_{cont\text{-}cat\text{-}H2O2} = 0.28 – 0.1 – 2\ g \cdot L^{-1}$ T = 16 ºC P = 1 atm pH = 3	$X_{cont} = 85\%$ $X_{TOC} < 30\%$ (t = 3 h)	After 5 rounds of recycle, the MNPs still remained almost 100% catalytic activity	[31]
Ferromagnetic nanoparticles (MNPs)	Phenol Aniline	$C_{cont\text{-}cat\text{-}H2O2} = 0.05 – 5 – 40\ g \cdot L^{-1}$ T = 35 ºC P = 1 atm pH = 7	$X_{phenol} = 100\%$ $X_{aniline} = 100\%$ $X_{TOC(phenol)} = 43\%$ $X_{TOC(aniline)} = 40\%$ (t = 6 h)	After 8 runs, X_{phenol} was reduced by 20% and X_{TOC} by 30%. Reasons: 1. Conglomeration of NMPs 2. Part of MNPs are rinsed during discarding of supernatants	[40]
Ferromagnetic nanoparticles (MNPs)	Rhodamine B	$C_{cont\text{-}cat\text{-}H2O2} = 0.01 – 0.6 – 1.4\ g \cdot L^{-1}$ T = 40 ºC P = 1 atm pH = 5.4 (Fe^{2+}/Fe^{3+} 1:1 ratio in synthesis)	$X_{cont} = 90\%$ (t = 1 h)	Iron leaching < 0.015 $mg \cdot L^{-1}$	[41]

Table 1. (Continued)

Catalyst	Pollutant	Operating conditions	Results	Stability	Reference
Ferromagnetic nanoparticles (MNPs) (Sono-Fenton)	Bisphenol A	$C_{cont\text{-}cat\text{-}H2O2} = 0.02 – 0.6 – 5.4\ g·L^{-1}$ T = 35 °C P = 1 atm pH = 5 (Ultrasound at 40 kHz, 100 w)	$X_{cont} = 100\%$ $X_{TOC} = 49\%$ (t = 8 h)	5 runs $X_{cont(5th)} = 70\%$ Reason: Loss of MNPs (0.5 mg·L^{-1}/run)	[42]
Nano-sized magnetic iron oxides	Phenol	$C_{cont\text{-}cat\text{-}H2O2} = 0.025 – 3 – 5\ g·L^{-1}$ T = 22 °C P = 1 atm pH = 7	$X_{cont} = 60\%$ (t = 24 h)	3 runs $X_{cont(1st)} = 60\%$ $X_{cont(2nd)} = 55\%$ $X_{cont(3rd)} = 30\%$ Iron leaching <0.01 mg·L^{-1}/run Reason: complexation of iron by organic acids (block active centers or cause a destruction of active surface structures)	[43]
Ferromagnetic nanoparticles (MNPs)	2,4-dichlorophenol	$C_{cont\text{-}cat\text{-}H2O2} = 0.1 – 1 – 0.4\ g·L^{-1}$ T = 30 °C P = 1 atm pH = 3	$X_{cont} = 100\%$ $X_{TOC} = 51\%$ (t = 3 h)	5 runs $X_{cont(1st)} = 95\%$ $X_{cont(3rd)} = 60\%$ $X_{cont(5th)} = 40\%$ (t = 2 h) Reasons: -Iron leaching (9.8 mg·L^{-1})	[44]

Catalyst	Pollutant	Operating conditions	Results	Stability	Reference
				-Reduction of catalyst specific area -Poisoning of active sites -Conglomeration of NMPs	
Maghemite nanoparticles (NP) and maghemite/ silica nanocomposite Microspheres (MS)	Methyl orange (MO) Methylene blue (MB) Paranitrophenol (PNP)	$C_{cont\text{-}cat\text{-}H2O2}$ = 0.08 (MO,MB), 0.035 (PNP) – 1.75 (Fe) – 34 $g \cdot L^{-1}$ T = 40 °C P = 1 atm pH = 6.5	$X_{NP\text{-}MO}$ = 100% $X_{NP\text{-}MB}$ = 82% $X_{NP\text{-}PNP}$ = 72% $X_{MS\text{-}MO}$ = 96% $X_{MS\text{-}MB}$ = 98% $X_{MS\text{-}PNP}$ = 67% (t = 4 h)	5 runs $X_{MS\text{-}MO(1st)}$ = 95% $X_{MS\text{-}MO(3rd)}$ = 92% $X_{MS\text{-}MO(5th)}$ = 90% Iron leaching = 7.3 $mg \cdot L^{-1}$	[45]
Fe^0/Fe_3O_4 composite	Methylene blue	$C_{cont\text{-}cat\text{-}H2O2}$ = 0.1 – 3 – 10 $g \cdot L^{-1}$ T = 25 °C P = 1 atm pH = 6	X_{TOC} = 75% (t = 2 h)	--	[46]
Magnetic mesoporous silica nanocomposite	Phenol	$C_{cont\text{-}cat\text{-}H2O2}$ = 0.2 – 5 – 1.7 $g \cdot L^{-1}$ T = 40 °C P = 1 atm pH = 4	X_{TOC} = 78% (t = 2 h)	3 runs $X_{TOC(1st)}$ = 78% $X_{TOC(3rd)}$ = 65% Iron leaching: 0.9 to 0.29 $mg \cdot L^{-1}$ from 1 to 3 run	[47]
Humic acid coated Fe_3O_4 magnetic nanoparticles	Sulfathiazole	$C_{cont\text{-}cat\text{-}H2O2}$ = 0.05 – 3 – 14 $g \cdot L^{-1}$ T = 40 °C P = 1 atm pH = 3.5	X_{cont} = 100% (t = 1 h) X_{TOC} > 90% (t = 6 h)	3 runs $X_{cont(1\text{-}3\ runs)}$ = 100% (t = 6 h) Iron leaching < 0.1 $mg \cdot L^{-1}$	[48]

Table 1. (Continued)

Catalyst	Pollutant	Operating conditions	Results	Stability	Reference
Ferromagnetic nanoparticles (MNPs) (Sono-Fenton)	Bisphenol A	$C_{cont\text{-}cat\text{-}H2O2}$ = 0.02 – 0.6 – 5.4 $g \cdot L^{-1}$ T = 35 °C P = 1 atm pH = 5 (Ultrasound at 40 kHz, 100 w)	X_{cont} = 100% X_{TOC} = 49% (t = 8 h)	5 runs $X_{cont(5th)}$ = 70% Reason: Loss of MNPs (0.5 $mg \cdot L^{-1}$/run)	[42]
Magnetic nanoscaled Fe_3O_4/CeO_2 composite	4-chlorophenol	$C_{cont\text{-}cat\text{-}H2O2}$ = 0.1 – 2 – 1 $g \cdot L^{-1}$ T = 30 °C P = 1 atm pH = 3	X_{cont} = 100% (t = 1 h)	6 runs $X_{cont(1st)}$ = 100% $X_{cont(3rd)}$ = 60% $X_{cont(6th)}$ = 40% (t = 1 h) High iron leaching = 12 $mg \cdot L^{-1}$	[49]
Sludge derived magnetic porous carbon	1-diazo-2-tnaphtol-4-sulfonic acid	$C_{cont\text{-}cat\text{-}H2O2}$ = 0.15 – 0.5 – 0.51 $g \cdot L^{-1}$ T = 25 °C P = 1 atm pH = 5	X_{cont} = 94% X_{TOC} = 48% (t = 2 h)	Iron leaching < 7.4 $ng \cdot L^{-1}$	[50]
Shape-controlled nanostructures magnetite-type materials	Phenol	$C_{cont\text{-}cat\text{-}H2O2}$ = 0.6 – 0.3 – 1.36 $g \cdot L^{-1}$ T = 25 °C P = 1 atm pH = 4	X_{cont} = 98% X_{TOC} = 74% (t = 1.5 h)	3 runs $X_{cont(1st)}$ = 98% $X_{cont(2nd)}$ = 97% $X_{cont(3rd)}$ = 93% $X_{TOC(1st)}$ = 74% $X_{TOC(2nd)}$ = 72% $X_{TOC(3rd)}$ = 68%	[51]

Catalyst	Pollutant	Operating conditions	Results	Stability	Reference
				Iron leaching $= 0.17\ mg \cdot L^{-1}/run$ Reasons: -partial oxidation of Fe^{2+} to Fe^{3+}	
Magnetic core-shell structural γ-Fe_2O_3@Cu/Al-MCM-41 nanocomposite	Phenol	$C_{cont\text{-}cat\text{-}H2O2} = 0.08 – 1 – 1.7\ g \cdot L^{-1}$ T = 40 °C P = 1 atm pH = 4	$X_{TOC} = 80\%$ (t = 2 h)	3 runs $X_{TOC(1st)} = 80\%$ $X_{TOC(2nd)} = 60\%$ $X_{TOC(3rd)} = 40\%$ Leaching: $Cu^{2+} = 0.18\ mg \cdot L^{-1}$ $Fe^{3+} = 0.12\ mg \cdot L^{-1}$	[52]
Magnetic core/shell structured g-Fe_2O_3@Ti-tmSiO_2	Methylene blue	$C_{cont\text{-}cat\text{-}H2O2} = 0.05 – 0.05 – 15\ g \cdot L^{-1}$ T = 25 °C P = 1 atm pH = 4	$X_{cont} = 99\%$ (t = 5.5 h)	6 runs No loss of activity Iron leaching = 0.17 $mg \cdot L^{-1}$	[53]
Magnetic composite nanospheres	Phenol	$C_{cont\text{-}cat\text{-}H2O2} = 0.21 – 0.1 – 1.36\ g \cdot L^{-1}$ T = 20 °C P = 1 atm pH = 5	$X_{cont} = 98\%$ $X_{COD} = 76\%$ (t = 2 h)	3 runs $X_{cont(1st)} = 98\%$ $X_{COD(1st)} = 76\%$ $X_{cont(3rd)} = 95\%$ $X_{COD(3rd)} = 70\%$ Iron leaching = 3%	[54]
Graphene oxide-Fe_3O_4 nanocomposites	Acid orange 7	$C_{cont\text{-}cat\text{-}H2O2} = 0.035 – 0.1 – 0.75\ g \cdot L^{-1}$ T = 25 °C P = 1 atm pH = 3	$X_{cont} = 100\%$ (t = 3 h)	--	[55]

Table 1. (Continued)

Catalyst	Pollutant	Operating conditions	Results	Stability	Reference
Magnetic ordered mesoporous copper ferrite	Imidacloprid	$C_{cont-cat-H2O2} = 0.01 – 0.3 – 1.36\ g \cdot L^{-1}$ T = 30 °C P = 1 atm pH = 3	$X_{cont} = 100\%$ $X_{TOC} = 33\%$ (t = 5 h)	5 runs $X_{cont(1st)} = 100\%$ $X_{cont(5th)} = 100\%$ Iron leaching $<1\ mg \cdot L^{-1}$	[56]

C = Concentration, X = Conversion, Cont = Contaminant, Cat = Catalyst, Toc = Total Organic Carbon, Cod = Chemical Oxygen Demand.

However, the degradation of the organic pollutants by iron-oxide minerals is directly related to the iron dissolution from the solid, which represents the key parameter for effective performance of the system [30, 32, 58, 59, 63]. Consequently, the iron mineral is used as dissolved iron source and the oxidation proceeds mostly upon homogenous Fenton. Moreover, the decomposition rate of organic pollutants is considerably slower than in classic Fenton oxidation where Fe(II) salts are used. Xue et al. (2009)[32] studied the degradation of pentachlorophenol using magnetite as catalyst under ambient conditions. They analyzed the effect of adding chelating agents (EDTA, CMCD, oxalate, tartrate, citrate and succinate), concluding that their presence in the reaction medium improved the dissolution of iron from the solid surface and then the occurrence of homogeneous Fenton reaction. Although almost complete conversion of pentachlorophenol was reached, 9 hours of reaction were required and a high concentration of dissolved iron (14 $mg \cdot L^{-1}$) was found. In the same way, Matta et al. (2008)[64] found that the addition of chelating agents (EDTA, CMCD) improved significantly the conversion of 2,4,6-trinitrophenol in the presence of magnetite from 25% to 50% and 62%, respectively.

In this context, searching active and stable minerals capable of accomplishing the oxidation with negligible iron leaching and at higher degradation rates is imperative. Liang et al. (2010)[33] established that natural vanadium-titanium magnetite is a mineral showing a high activity and stability in the oxidation of Acid Orange II under mild conditions. In that work, the degradation process was dominated by the heterogeneous Fenton oxidation and complete conversion of the pollutant was achieved in 4 h reaction time. The small decrease of catalytic activity upon three consecutive runs was due to slight leaching of iron from the catalyst. Consistent with these results, Zhong et al. (2012)[34] also found that titanomagnetite retained the catalytic activity after three cycles in UV/Fenton oxidation of Tetrabromobisphenol-A, suggesting a good stability and reusability.

Although the most recent results have shown the good stability of some magnetic minerals, the degradation rate of the pollutants is still far from that obtained upon homogeneous Fenton oxidation under similar operating conditions [21]. In this sense, improving the activity of the catalysts and the rate of the process is essential. Following this objective, some authors have studied the effect of including other cations into the magnetite structure by isomorphic substitution of Fe. Costa et al. (2006)[65] analyzed the effect of incorporating Mn, Co and Ni into the magnetite structure. They found that the presence of Co or Mn produced a remarkable increase of the activity, whereas Ni inhibited the H_2O_2 decomposition in CWPO of methylene blue. Similarly, Baldrian et al. (2006)[66] showed that magnetite doped with Co, Cu or Mn allowed the effective oxidation of various synthetic dyes, maintaining its activity upon sequential runs. The presence of Mn and Co cations seems to promote the decomposition of H_2O_2 into ·OH radicals and accelerate electron transfer giving rise to a more efficient regeneration of Fe(II). However, the main advantage of using natural magnetic minerals is that they can be used directly, without any modification and thus, maintaining the low cost of those materials. In this sense, another possibility to improve the catalytic performance could be the intensification of the process. So far, all the studies related to the use of natural minerals have been carried out at ambient temperature. However, the recent findings in Fenton oxidation have shown that increasing the temperature leads to a more efficient consumption of H_2O_2 which indicates an enhanced iron-catalyzed H_2O_2 decomposition into radicals, thus clearly improving the oxidation rate and mineralization degree [10, 21]. In fact, some commercial applications of the Fenton process at industrial scale are carried out at 120 ºC (OHP®, MFC-Foret®). Accordingly, we have recently studied the effect of increasing the temperature (25-90 ºC) in the decomposition of H_2O_2 catalyzed by three natural minerals (hematite, magnetite and ilmenite). Ilmenite is a crystalline iron titanium oxide ($FeTiO_3$) where Fe and Ti occupy alternating layers. Due to its magnetic properties and low cost it could also represent an interesting catalyst for CWPO. However, it has not been previously used for that purpose due probably to its low rate of H_2O_2 decomposition [67]. So far, it has been successfully tested for the degradation of various organic pollutants by photocatalysis [67, 68].

As can be seen in Figure 6, the decomposition of H_2O_2 was significantly improved at increasing temperature. At ambient temperature, none of the minerals tested showed a significant rate of H_2O_2 decomposition, which is consistent with previous results reported in the literature. According to our results, increasing the temperature represents an interesting alternative to improve the rate of the process. However, it has to be taken into account that appreciable amounts of dissolved iron were measured in the reaction medium in the case of magnetite (4 $mg \cdot L^{-1}$ at 75 ºC) whereas hematite and ilmenite were fairly stable (0.2 and 0.5 $mg \cdot L^{-1}$, respectively).

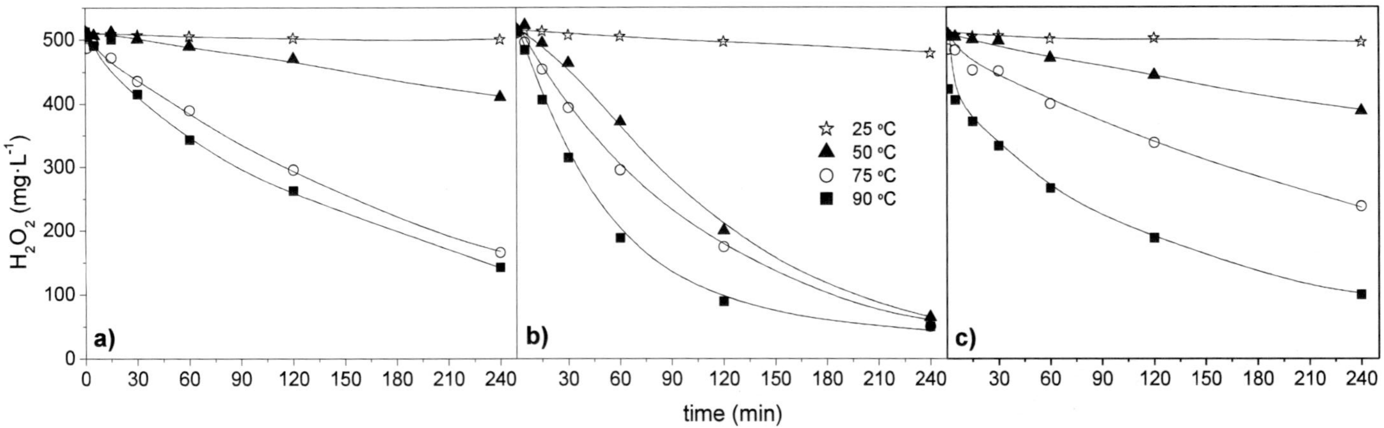

Figure 6. H_2O_2 decomposition catalyzed by hematite (a), magnetite (b) and ilmenite (c) natural minerals ($[mineral]_0 = 1\ g \cdot L^{-1}$; $[H_2O_2]_0 = 500\ mg \cdot L^{-1}$; $pH_0 = 3$).

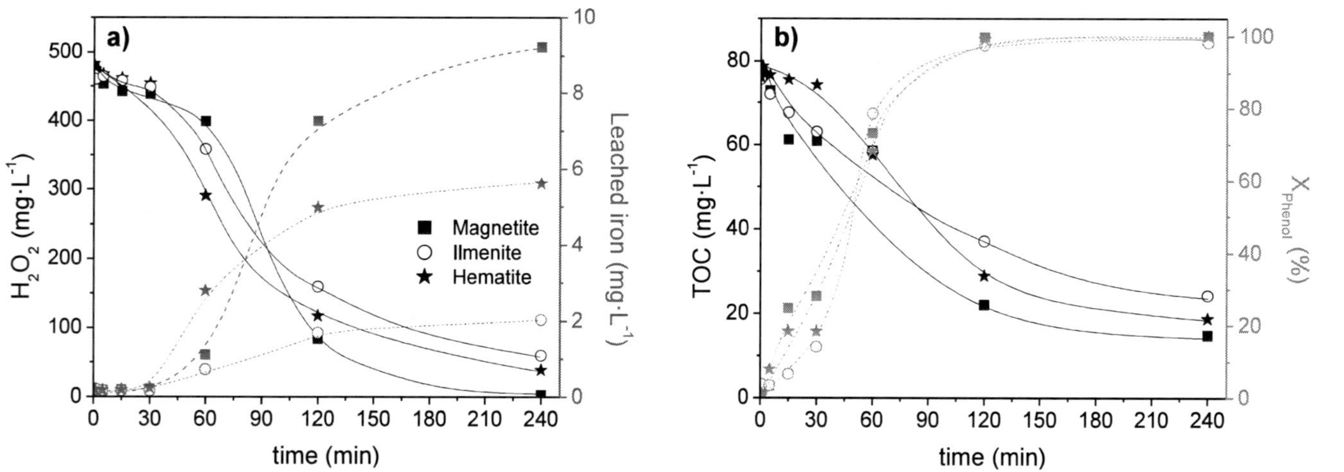

Figure 7. Evolution of phenol conversion, TOC, H_2O_2 decomposition and iron leaching in CWPO with magnetite, ilmenite and hematite minerals as catalysts ($[phenol]_0 = 100$ mg·L^{-1}; $[H_2O_2]_0 = 500$ mg·L^{-1}; $[mineral]_0 = 1$ g·L^{-1}; T = 75 ºC; $pH_0 = 3$).

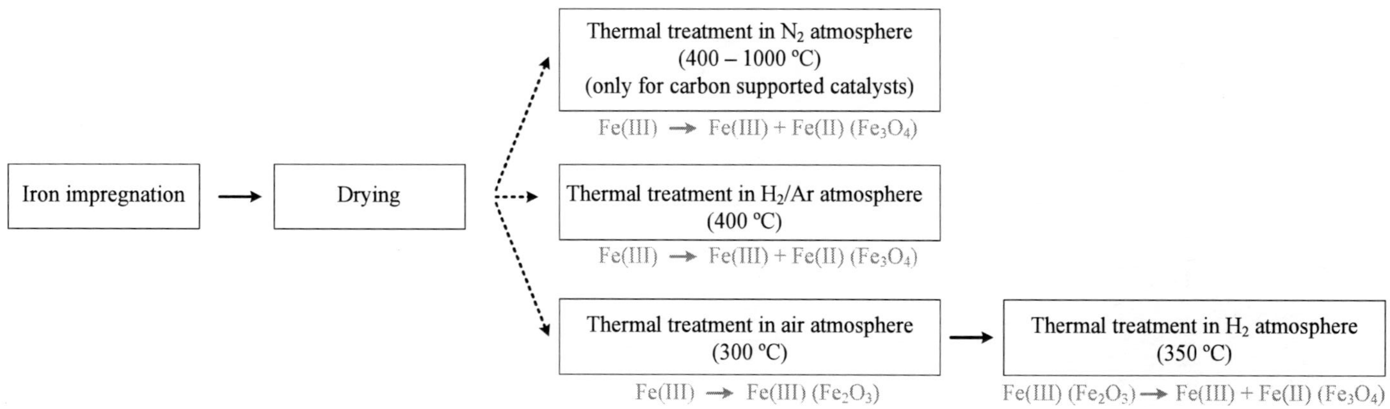

Figure 8. Preparation of magnetic catalysts by *in situ* synthesis of Fe_3O_4.

To gain further insight on the catalytic performance of natural minerals, the oxidation of phenol was carried out at 75 °C, paying special attention to iron leaching. The results obtained are depicted in Figure 7. The three minerals allowed achieving complete conversion of phenol in 2 h reaction time with high degrees of mineralization, above 60% at the end of the experiments (4 h). Nevertheless, iron leaching was especially high from magnetite, the oxidation process taking place in that case predominantly by homogeneous reaction. It has to be highlighted the potential of ilmenite as CWPO catalyst since it showed a considerable resistance to iron leaching. Moreover, the mineralization degree achieved is even higher than the obtained upon homogeneous Fenton oxidation under similar operating conditions [21].

Preparation of Magnetic Catalysts by *in Situ* Synthesis of Magnetite

In the last few years, the development of magnetic catalysts by *in situ* synthesis of magnetite using different thermal treatments is gaining attention. In contrast to the complexity, high cost and difficult scale-up of production of magnetic nanoparticles, conventional methods represent a simpler and effective way. Briefly, an aqueous solution of the iron salt is deposited on the support, which is subsequently subjected to different thermal treatments in order to develop the formation of magnetite, and thereby, the appearance of magnetic properties in the solid. Figure 8 shows the main steps followed in the preparation of magnetic catalysts by *in situ* synthesis of Fe_3O_4.

The supports used so far for the preparation of this kind of magnetic catalyst have been activated carbon and γ-Al_2O_3 (Table 1), although their preparation could be extended to other materials traditionally used in CWPO such as pillared clays, silica or zeolites.

Activated carbon has proved to be an excellent catalytic support in CWPO due to its favorable properties such as high surface area, well developed and different porous structures, variable surface composition, good chemical resistance and relatively low cost [23, 24]. Nguyen et al. (2011)[36] prepared magnetic carbon catalyst by wetness impregnation of activated carbon with $Fe(NO_3)_3 \cdot 9H_2O$. The impregnated solid was dried at room temperature and further treated at 600 °C for 1 h under N_2 atmosphere to produce magnetite.

XRD analyses of the catalyst confirmed the presence of magnetite whereas its magnetic properties were further characterized by magnetic measurements, obtaining a saturation magnetization value (M_S) of 6 emu·g^{-1}. These results showed that the catalyst can be manipulated by an external magnetic field, such as magnet. The catalyst (2.5 g·L^{-1}) was tested in the CWPO of methyl orange (50 mg·L^{-1}) at 30 °C and pH 4. Complete conversion of this compound and close to 60% mineralization were achieved after 2 h reaction time. The stability of the catalyst was demonstrated in three consecutive experiments with very low iron leaching, so that the reaction is essentially heterogeneous. A slight loss of activity was observed, attributed to the small iron leaching.

Chun et al. (2012)[37] synthesized magnetite mesocellular carbon foam (MSU-F-C) by wet impregnation using an ethanolic solution of $Fe(NO_3)_3·9H_2O$ as iron precursor. After evaporation of the ethanol under stirring, the resulting material was heated to 400 °C for 4 h under a H_2/Ar atmosphere. The XRD peaks were indexed to the Fe_3O_4 crystal structure and broad diffraction peaks corresponding to the amorphous carbon support. The hysteresis loop of the Fe_3O_4/MSU-F-C showed a high M_S (32 emu·g^{-1}). Thus, it can be easily separated with a magnet and re-dispersed into solution quickly without severe aggregation. An average particle size of 15 nm for Fe_3O_4 was reported. The catalyst (0.1 g·L^{-1}) showed a high activity and stability in CWPO of phenol (10 mg·L^{-1}) at ambient conditions and pH 3. Almost complete conversion of phenol (95%) was achieved after 4 h reaction time. The homogenous Fenton contribution due to dissolved species from the catalyst was insignificant (<0.6 mg·L^{-1}) and non-appreciable loss of activity was observed after multiple consecutive runs.

Tristao et al. (2014)[39] synthesized a carbon matrix containing Fe_3O_4 particles and tested it in CWPO of methylene blue. For the synthesis of the catalyst, aqueous solutions of $Fe(NO_3)_3·9H_2O$ containing 1, 4, and 8 wt.-% Fe at pH 1 (HCl) were heated and commercial sucrose was added. The solution was evaporated and thermally treated at 400, 600 and 800 °C in N_2 atmosphere for 1 h. XRD analyses confirmed the presence of Fe_3O_4 with crystallite sizes between 13 and 20 nm in the catalyst treated at 400 °C, while those treated at higher temperatures contained metallic iron and Fe_3C. Methylene blue (200 mg·L^{-1}) was almost completely removed (95%) after 3 h reaction time using 4.3 g·L^{-1} of catalyst (8%Fe-400) at room temperature and pH 6.

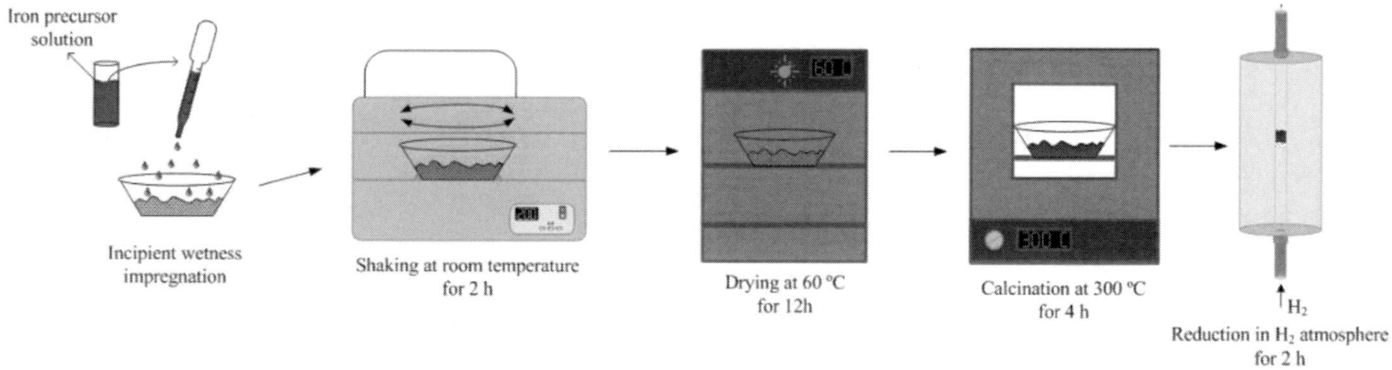

Figure 9. Method of preparation of ferromagnetic Fe_3O_4/γ-Al_2O_3 catalyst. Adapted from [69].

Gu et al. (2012)[38] synthesized a magnetic porous carbon containing Fe_3O_4 (FPC) by carbonization and activation of sewage sludge without extra addition of ferric ions, taking advantage of its high iron content (≈30%). The solids were subjected to pyrolysis at different temperatures (600, 800 and 1000 ºC) in N_2 atmosphere for 2 h. The FPC600 showed a well-developed porous structure, with mesopores much wider than those of FPC800 and FPC1000. Thus, it was selected as optimum for CWPO experiments. The degradation of a dye (1-diazo-2-naphtol-4-sulfonic acid) (250 $mg \cdot L^{-1}$) was studied using 0.5 $g \cdot L^{-1}$ catalyst under mild conditions at pH 5. Almost complete conversion of 1-diazo-2-naphtol-4-sulfonic acid (97%) and a high mineralization degree (87%) were achieved after 4 h reaction time. The M_s value was 8.7 $emu \cdot g^{-1}$, while the XRD pattern revealed the presence of Fe_3O_4, Fe_3C and α-Fe. The catalyst showed a good stability after three sequential applications although a slight loss of activity was appreciated due to some iron leaching (0.77% of the total iron content).

Alumina (Al_2O_3) is widely used as catalyst support due to its mechanical, electrical and chemical stability as well as its relatively low cost. So far, the catalysts based on Al_2O_3 instead of activated carbon have shown a considerably higher stability upon CWPO [26]. Bautista et al. [26, 28] prepared a $Fe_2O_3/\gamma\text{-}Al_2O_3$ catalyst and applied it in the CWPO of cosmetic and phenolic wastewaters. The catalyst showed remarkable stability in long-term continuous experiments with limited iron leaching. More recently, we added a reduction stage in H_2 atmosphere to the preparation procedure in order to reduce Fe_2O_3 to Fe_3O_4 and thus, developing magnetic properties in the catalyst [27]. The activity and stability of this new magnetic catalyst were compared with those of a conventional one with the same support. The catalysts were prepared by incipient wetness impregnation of powdered $\gamma\text{-}Al_2O_3$ with a $Fe(NO_3)_3 \cdot 9H_2O$ aqueous solution, adjusting the Fe load to a nominal 4 wt.-%. After impregnation, the solids were left for 2 h at room temperature, dried for 12 h at 60 ºC and calcined for 4 h at 300 ºC. In order to develop magnetic properties in one of the catalysts, it was subjected to reduction in H_2 atmosphere for 2 h at 350 ºC. Figure 9 shows the synthesis procedure. XRD, XPS and Mössbauer spectroscopy analyses confirmed the presence of hematite (Fe_2O_3) in the conventional catalyst and magnetite (Fe_3O_4) in the new one. The iron nanoparticles presented a wide range of sizes within 5-45 nm, showing a mean diameter of 25 and 33 nm for $Fe_2O_3/\gamma\text{-}Al_2O_3$ and $Fe_3O_4/\gamma\text{-}Al_2O_3$, respectively. The M_S value was 2.24 emu·g^{-1}, which allowed easy separation of the solid from the reaction mixture by applying an external magnetic field, such as a magnet (Figure 10). Both catalysts (1 g·L^{-1}) were applied to the CWPO of 4-chlorophenol, 2,4-dichlorophenol and 2,4,6-trichlorophenol (100 mg·L^{-1}) at 50 °C and pH 3. Complete dechlorination of all the chlorophenols tested was achieved with mineralization degrees above 70% after 4 h reaction time. The residual by-products were short-chain organic acids without significance in terms of ectoxicity. The magnetic catalyst decomposed H_2O_2 about three times faster than the conventional one (7.3 x 10^{-3} and 2.4 x 10^{-3}min^{-1}, respectively). Thus, the degradation rate of the target pollutants was also faster. Both catalysts showed a high stability upon long-term continuous experiments (100 h). As can be seen in Figure 11, H_2O_2 and chlorophenol conversions remained almost constant over the 100 h on stream. The loss of iron from the catalyst was around 5% at the end of the experiment. The BET surface remained unchanged and only small amounts of residual carbon-containing species (≈1.2 wt.-%) were measured on the catalyst surface.

Although the magnetic catalysts developed so far have shown to be fairly active and stable in CWPO of different organic compounds, the oxidation rates were far below that the obtained by conventional homogenous Fenton oxidation. With the aim of improving the rate of the process, we have recently explored the possibility of intensifying the process by increasing the temperature (50 to 90 °C). The results have been compared with those obtained under homogeneous Fenton oxidation [70]. The magnetic catalyst (1 g·L^{-1}) allowed achieving 90% TOC reduction after 1 h reaction time at 100 mg·L^{-1} chlorophenol initial concentration, pH 3 and 90 °C temperature. That degree of mineralization is considerably higher than the obtained upon Fenton oxidation (60%) under the same operating conditions (Figure 12). The leaching of iron from the catalyst was even lower than the observed at 50 °C (2.5 and 0.5 mg·L^{-1} Fe^{3+} at 50 and 90 °C, respectively) because the concentration of oxalic acid was significantly reduced. This acid has proved to provoke a negative effect on the stability of supported Fe catalysts due to metal leaching [23]. The catalyst showed a remarkable stability upon three sequential runs without significant loss of activity and was easily separated from the reaction medium because advantage of its magnetic properties. Thus, the use of the magnetic $Fe_3O_4/\gamma\text{-}Al_2O_3$ catalyst at high temperature (90 °C) clearly represents an interesting alternative to homogeneous Fenton oxidation.

Figure 10. Illustration of the magnetic recovery of the Fe_3O_4/γ-Al_2O_3 catalyst from the reaction medium.

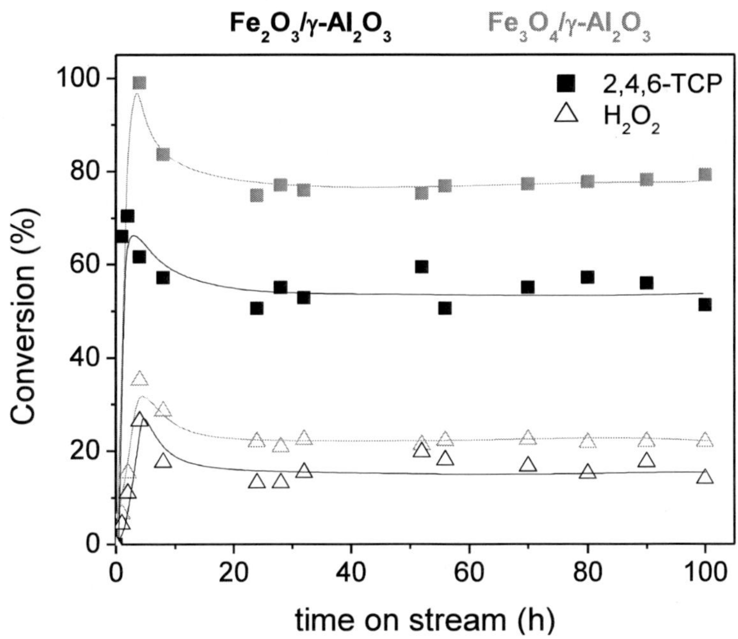

Figure 11. Long-term performance of γ-Al_2O_3-supported hematite and magnetite catalysts upon CWPO of 2,4,6-TCP ($[catalyst]_0 = 1\ g{\cdot}L^{-1}$; $[2,4,6\text{-trichlorophenol}]_0 = 100\ mg{\cdot}L^{-1}$; $[H_2O_2]_0 = 180\ mg{\cdot}L^{-1}$; T = 50 ºC; $pH_0 = 3$; $\tau = 16.5$ kgcat·h·mol^{-1}).

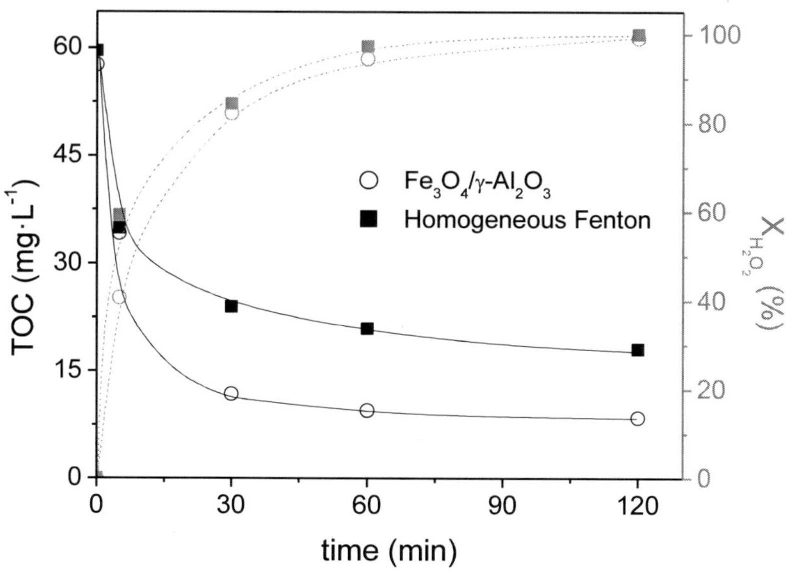

Figure 12. TOC reduction and H_2O_2 decomposition upon 4-CP degradation by CWPO and homogeneous Fenton oxidation ($[Fe_3O_4/\gamma\text{-}Al_2O_3]_0 = 1\ g\cdot L^{-1}$; $[Fe^{3+}]_0 = 10\ mg\cdot L^{-1}$; $[\text{4-chlorophenol}]_0 = 100\ mg\cdot L^{-1}$; $[H_2O_2]_0 = 350\ mg\cdot L^{-1}$; T = 90 °C; $pH_0 = 3$).

Ferromagnetic Nanoparticles as Fenton Catalysts

Magnetic Fe_3O_4 nanoparticles (MNPs) have been reported as catalyst in many reactions such as Fischer-Tropsch, Haber-Bosch, environmental catalysis and peroxidase-like activities [55]. Recently, a new generation of heterogenous catalysts based on MNPs has been developed for Fenton oxidation. The importance of nanoparticles is related to their inherent properties, which can differ from those of the macroscopic or bulk forms of the same material [71]. Furthermore, the activity and selectivity of nanocatalysts are strongly dependent on their size, shape and surface structure, as well as on their bulk composition.

Different methods have been studied for the synthesis of MNPs, including co-precipitation [31, 72], microemulsion [73], hydrothermal [51, 74] or sonochemical synthesis [41], among other [71]. The most conventional method is, by far, co-precipitation [71, 75]. It consists of mixing ferric to ferrous ions in a 1:2 molar ratio in highly basic solutions at room or higher temperature. The size and shape of MNPs can be modified by selecting the type of salt, the ferric and ferrous ions ratio, the reaction temperature, the pH value and the ionic strength of the medium [75]. The main disadvantage of this method is related to the wide particle size distribution achieved, which usually requires secondary size selection. A microemulsion is a thermodynamically stable isotropic dispersion of two immiscible phases, commonly water and oil, under presence of a surfactant. Self-assembled structures such as spherical micelles or lamellar phases can be formed, thus shape- and size- controlled MNPs can be synthesized by this method [75]. Nevertheless, aggregation of the NMPs usually needs several washing processes and further stabilization treatments. Hydrothermal synthesis includes various wet chemical technologies of crystallizing the substance involving high temperature (130 to 250 ºC) at high pressure (0.3 to 4 MPa). In sonochemical synthesis, MNPs are formed by the implosive collapse of bubbles through acoustic cavitation. The extreme conditions formed in these hotspots (5000 K, pressure of 1800 atm and cooling rates in excess of 10^{10} $K \cdot s^{-1}$) are beneficial for the formation of MNPs and have a shear effect for agglomeration [75]. Thermal decomposition is based on the decomposition of an organic solution phase ($Fe(cup)_3$ (cup = N-nitrosophenylhydroxylamine), $Fe(acac)_3$ (acac = acetylacetonate) or $Fe(CO)_5$) followed by oxidation leading to high-quality monodispersed MNPs. However, it requires relatively high temperatures and a complicated operation.

Zhang et al. (2008) [31] studied the degradation of phenol by CWPO using MNPs as catalyst. MNPs were prepared by co-precipitation and appeared approximately spherical with an average diameter of 13 nm. The M_S value was 65 $emu \cdot g^{-1}$, indicating their magnetic character. Oxidation runs were carried out at 16 ºC and pH 3, using an initial concentration of phenol of 280 $mg \cdot L^{-1}$, 100 $mg \cdot L^{-1}$ NMPs and 2 $g \cdot L^{-1}$ H_2O_2. After 3 h reaction time, phenol reached 85% although the mineralization degree did not exceed 30%. The reusability of MNPs was assessed by using the same MNPs regenerated simply by sonication and washing with deionized water. After 5 rounds of recycle, the MNPs maintained almost invariable their catalytic activity. Zhang et al. (2009)[40] used MNPs, prepared by co-precipitation, as catalyst for CWPO of phenolic and aniline compounds. The experiments were performed at 35 ºC at neutral pH using 5 $g \cdot L^{-1}$ MNPs, 40 $g \cdot L^{-1}$ H_2O_2 and 50 $mg \cdot L^{-1}$ pollutant concentration. After 6 h of reaction, complete conversion of phenol and aniline were achieved, with a TOC reduction around 40%. After 8 sequential uses of MNPs, conversion of phenol and TOC were reduced by 20 and 30%, respectively. The main reasons for the loss of activity were attributed to the conglomeration of MNPs and to the fact that MNPs might be rinsed during discarding of supernatants. Similar results were obtained by

Wang et al. (2010) [41] upon CWPO of Rhodamine B (10 mg·L^{-1}) by MNPs (600 mg·L^{-1}) at 40 °C and pH 5. In 1 hour reaction time 90% of the pollutant was removed with limited iron leaching.

More recently, Xu and Wang (2012a) [44] applied MNPs, prepared by co-precipitation, in CWPO of 2,4-dichlorophenol at ambient conditions (30 °C, pH 3, 100 mg·L^{-1} 2,4-dichlorophenol, 1 g·L^{-1} MNPs, 0.4 g·L^{-1} H_2O_2), achieving good results in terms of activity (complete conversion of the pollutant and TOC reduction above 50% after 3 h reaction time). However, the stability of MNPs was not satisfactory, since the pollutant conversion decreased from 95% to 40% after 2 h. The main reason for that loss of activity was iron leaching (9.8 mg·L^{-1}), but also other factors such as reduction of the catalyst surface area, poisoning of the active catalytic sites by adsorbed organic species, aggregation of MNPs, and the discarding of supernatants during the rinsing of MNPs. The lack of stability of MNPs upon CWPO has been also claimed by other authors. Huang et al. (2012) [42] showed a 30% reduction of activity in CWPO of Bisphenol A after 5 successive runs. Similarly, Rusevova et al. (2012) [43] found a decrease of 50% on the catalytic activity of MNPs upon 3 sequential uses. The authors attributed it to the complexation of iron by organic acids, which block active centers or cause a destruction of active surface structure. Likewise, [45] Ferroudj et al. (2013) [45] attributed the loss of activity upon sequential application of magnetic maghemite nanoparticles to iron leaching (7.3 mg·L^{-1}).

The main shortcomings related to the use of unsupported MNPs as CWPO catalysts are that they are prone to aggregate, thus losing their dispersibility and reducing their activity [55]. Their lack of reusability is due to iron leaching [44]. Moreover, they can be rinsed during discarding supernatants [40]. In this sense, the immobilization of MNPs onto high-surface-area supports represents a solution which would preserve their unique properties.

Xia et al. (2011) [47] synthesized a magnetically separable mesoporous silica nanocomposite and applied it to phenol oxidation. The catalyst presented highly dispersed iron species, thus providing enough active sites for CWPO, exhibiting a remarkable catalytic performance. Oxidation of phenol (200 mg·L^{-1}) was carried out using 5 g·L^{-1} catalyst and 1.7 g·L^{-1} H_2O_2 at 40 °C and pH 4. Phenol conversion was around 80% after 2 h reaction time, although it was reduced to 65% after reusing the catalyst for the third time. The reason was mainly attributed to iron leaching (1 mg·L^{-1}). Xu and Wang (2012b) [49] also found a loss of activity upon 6 runs (from 100% to 40% removal for 1st to 6th runs, respectively) for 4-chlorophenol oxidation with Fe3O4/CeO2 composite. In this case, the iron leaching was around 12 mg·L-1, equivalently about 1.9% of the total iron present in the catalyst. Likewise, Hou et al. (2014) [51] evidenced a loss of activity (≈10% upon 3 succesive runs) of shape-controlled nanostructures of magnetite-type materials upon phenol CWPO at ambient conditions. They attributed the decrease on catalytic activity to partial oxidation of Fe(II) to Fe(III) (the Fe(II)/Fe(III) decreased to 41% from the 53% of the fresh material). Ling et al. (2014) [52] also observed the low stability of magnetic core-shell structural γ-Fe_2O_3@Cu/Al-MCM-41 nanocomposite upon successive application to phenol oxidation.

On the contrary, Niu et al. (2011) [48] established that humic acid-coated MNPs present a remarkable stability in CWPO of sulfathiazole, maintaining complete conversion upon 3 successive runs. However, those authors did not provide information about TOC reduction upon sequential cycles. Lv et al. (2014) [53] demonstrated the high stability of magnetic core-shell structured γ-Fe_2O_3@Ti-tmSiO_2 upon methylene blue oxidation in 6 successive runs showing negligible iron leaching. In the same line, sludge-derived magnetic porous carbon, graphene oxide-Fe_3O_4 nanocomposites and magnetic ordered mesoporous copper ferrite have also shown a remarkable stability with low iron leaching in CWPO of different organic pollutants at ambient conditions [50, 55, 56].

Conclusion

Magnetite-based catalysts represent an interesting alternative to homogeneous Fenton oxidation as well as to the conventional catalysts used in CWPO. Magnetic natural minerals of high availability and low cost, are gaining considerable attention for that purpose. However, although they have shown a high activity in CWPO due to the presence of Fe(II) in their structure, they suffer from iron leaching which limits their reusability. The *in situ* formation of magnetite in Fe-supported catalysts by thermal treatments represents also an interesting option for CWPO. These catalysts have shown to be highly active and remarkably stable, although the rate of oxidation is considerably lower than the obtained by the homogeneous Fenton process.

MNPs are presented as powerful catalysts. Their activity and selectivity are strongly dependent on their size, shape and surface structure, as well as on their bulk composition, which can be adequately tuned during their synthesis. Due to their excellent properties, MNPs have gained increased attention in the last five years and currently are the most studied magnetic materials for application in CWPO. However, their use is limited due to a number of shortcomings such as aggregation and subsequent loss of dipersibility and iron leaching. Moreover, they can be rinsed during discarding supernatants. Although their immobilization onto a support could be a potential solution, the attempts reported so far showed that iron leaching cannot be completely avoided. Thus, there is room for considerable improvements in this field. The major challenges point to the use of increasingly smaller sizes of nanoparticles since it can enhance the activity and stability of the particles as well as stabilize the nanoparticles against aggregation.

In general, the results obtained so far with MNPs as CWPO catalyst did not substantially improve those obtained with natural magnetic minerals or catalysts prepared by *in situ* synthesis of magnetite. In fact, the works carried out using nanoparticles have been characterized by the use of high concentrations of catalyst and H_2O_2 doses, much higher than the stoichiometric amounts. Moreover, although cost-effective manufacture is clearly stated in the cases of magnetic natural minerals and conventional catalysts with *in situ* synthesis of magnetite, it must be still demonstrated with the MNPs to allow their widespread application in wet peroxide oxidation.

Research on the potential application of magnetite-based catalysts in CWPO is still incipient but deactivation due to iron leaching is identified as one important limitation and the kinetics of the process should be improved trying to approach to homogeneous Fenton oxidation in order to be competitive. This scenario opens the door to the intensification of the process by increasing the temperature, alternative which has not been widely explored in the literature so far but has shown promising results, as stated in this chapter. The kinetics, degree of mineralization and stability of the catalysts can be considerably improved upon intensification of the process by increasing the reaction temperature.

ACKNOWLEDGMENTS

The authors acknowledge the Spanish MICINN through the projects CTQ2008-03988 and CTQ2010-14807 and by the CM through the project S-2009/AMB-1588.

REFERENCES

[1] Fenton, H. J. H. (1894). *J. Chem. Soc.*, *65*, 899.

[2] Eisenhauer, H. R. (1964). *J. Water Pollut. Control Fed.*, *36*, 1116.

[3] Pignatello, J. J., Oliveros, E. & MacKay, A. (2006). *Crit. Rev. Env. Sci. Technol.*, *36*, 1.

[4] Andreozzi, R., Caprio, V., Insola, A. & Marotta, R. (1999). *Catal. Today*, *53*,.

[5] Esplugas, S., Giménez, J., Contreras, S., Pascual, E. & Rodríguez, M. (2002). *Water Res.*, *36*, 1034.

[6] Perez, M., Torrades, F., Domenech, X. & Peral, J. (2002). *J. Chem. Technol. Biotechnol.*, *77*, 525.

[7] Azbar, N., Yonar, T. & Kestioglu, K. (2004). *Chemosphere*, *55*, 35.

[8] Cañizares, P., Paz, R., Sáez, C. & Rodrigo, M. A. (2009). *J. Environ. Manage.*, *90*, 410.

[9] Bautista, P., Mohedano, A. F., Gilarranz, M. A., Casas, J. A. & Rodriguez, J. J. (2007). *J. Hazard. Mater.*, *143*, 128.

[10] Pliego, G., Zazo, J. A., Blasco, S., Casas, J. A. & Rodriguez, J. J. (2012). *Ind. Eng. Chem. Res.* 51. 2888-2896.

[11] Rivas, F. J., Beltran, F. J., Gimeno, O. & Frades, J. (2001). *J. Agric. Food Chem.* 49. 1873-1880.

[12] Lucas, M. S. & Peres, J. A. (2009). *J. Hazard. Mater.*, *168*, 1253.

[13] Pliego, G., Zazo, J. A., Casas, J. A. & Rodriguez, J. J. (2013). *J. Hazard. Mater.*, 252-253, 180.

[14] Munoz, M., Pliego, G., de Pedro, J. A., Casas, J. J. Rodriguez, (2014). *Chemosphere*, *109*, 34.

[15] Bautista, P., Mohedano, A. F., Casas, J. A., Zazo, J. A. & Rodriguez, J. J. (2008). *J. Chem. Technol. Biotechnol.*, *83*, 1323.

[16] Chen, R. & Pignatello, J. (1997). *Environ. Sci. Technol.*, *31*, 2399.

[17] Neyens, J. B. E. (2003). *J. Hazard. Mater. B98*, 33

[18] Duesterberg, C. K., Cooper, W. J. & Waite, T. D. *Environ. Sci. Technol.*, *39*, 5052. (2005).

[19] Pera-Titus, M., García-Molina, V. Baños, M. A., Giméneza, J. & Esplugas, S. (2004). *Appl. Catal., B.*, *47*, 219.

[20] Munoz, M., de Pedro, Z.M., Casas, J. A. & Rodriguez, J. J. (2011). *J. Hazard. Mater.*, *190*, 993.

[21] Zazo, J. A., Pliego, G., Blasco, S. & Casas, J. A. J.J. Rodriguez, (2011). *Ind. Eng. Chem. Res.*, *50*, 866.

[22] Ovejero, G., Sotelo, J. L., Martínez, F., Melero, J. A. & Gordo, L. (2001). *Ind. Eng. Chem. Res.*, *40*, 3921.

[23] Zazo, J. A., Casas, J. A., Mohedano, A. F. & Rodriguez, J. J. (2006). *Appl. Catal., B.*, *65*, 261.

[24] Rey, A., Faraldos, M., Casas, J. A., Zazo, J. A., Bahamonde, A. & Rodriguez, J. J. (2009). *Appl. Catal. B.*, *86*, 69.

[25] Satishkumar, G., Landau, M., Buzaglo, T., Frimet, L., Ferentz, M., Vidruk, R., Wagner, F., Gal, Y. & Herskowitz, M. (2013). *Appl. Catal., B.*, 138-139, 276.

[26] Bautista, P., Mohedano, A. F., Menéndez, N., Casas, J. A., Zazo, J. A. & Rodriguez, J. J. (2011). *J. Chem. Technol. Biotechnol.*, *86*, 497.

[27] Munoz, M., de Pedro, Z. M., Menendez, N., Casas, J. A. & Rodriguez, J. J. (2013). *Appl. Catal., B.*, *136-137*, 218.

[28] Bautista, P., Mohedano, A. F., Menéndez, N., Casas, J. A. & Rodriguez, J. J., (2010). *Catal. Today*, *151*, 148.

[29] Kwan, W. P. & Voelker, B. M. (2003). *Environ. Sci. Technol.*, *37*, 1150.

[30] Matta, R., Hanna, K. & Chiron, S. (2007). *Sci. Total Environ.*, *385*, 242.

[31] Zhang, J., Zhuang, J., Gao, L., Zhang, Y., Gu, N., Feng, J., Yang, D., Zhu, J. & Yan, X. (2008). *Chemosphere*, *73*, 1524.

[32] Xue, X., Hanna, K., Despas, C., Wu, F. & Deng, N. (2009). *J. Mol. Catal. A: Chem.*, *311*, 29.

[33] X. Liang, Y. Zhong, S. Zhu, J. Zhu, P. Yuan, H. He, J. Zhang, (2010). *J. Hazard. Mater.*, *181*, 112.

[34] Y. Zhong, X. Liang, Y. Zhong, J. Zhu, S. Zhu, P. Yuan, H. He, J. Zhang, (2012). *Water Res.*, *46*, 4633.

[35] M. Usman, P. Faure, K. Hanna, M. Abdelmoula, C. Ruby, (2012). *Fuel*, *96*, 270.

[36] Nguyen, T. D., Phan, N. H., Do, M. H. & Ngo, K. T. (2011). *J. Hazard. Mater.*, *185*, 653.

[37] Chun, J., Lee, H., Lee, S., Hong, S., Lee, J., Lee, C. & Lee, J. (2012). *Chemosphere*, *89*, 1230.

[38] Gu, L., Zhu, N., Zhou, P. (2012). *Bioresour. Technol.*, *118*, 638.

[39] Tristão, J. C., Gomes de Mendonça, F., Lago, R. M., Ardisson, J. D. (2014). *Env. Sci. Poll. Res.* doi: 10.1007/s11356-014-2554-z.

[40] Zhang, S., Zhao, X., Niu, H7., Shi, Y., Cai, Y. & Jiang, G. (2009). *J. Hazard. Mater.*, *167*, 560.

[41] Wang, N., Zhu, L., Wang, D., Wang, M., Lin, Z. & Tang, H. (2010). *Ultrason. Sonochem.*, *17*, 526.

[42] Huang, R., Fang, Z., Yan, X. & Cheng, W. (2012). *Chem. Eng. J.*, *197*, 242.
[43] Rusevova, K., Kopinke, F. & Georgi, A. (2012). *J. Hazard. Mater.*, 241–242, 433.
[44] Xu, L. & Wang, J. (2012). *Appl. Catal., B.*, 123-124, 117.
[45] Ferroudj, N., Nzimoto, J., Davidson, A., Talbot, D., Briot, E., Dupuis, V., Bée, A., Medjram, M. S. & Abramson, S. (2013). *Appl. Catal., B.*, *136–137*, 9.
[46] Costa, R. C. C., Moura, F. C. C., Ardisson, J. D., Fabris, J. D. & Lago, R. M. (2008). *Appl. Catal., B.*, *83*, 131.
[47] Xia, M., Chen, C., Long, M., Chen, C., Cai, W. & Zhou, B. (2011). *Microporous Mesoporous Mater.*, *145*, 217.
[48] Niu, H., Zhang, D., Zhang, S., Zhang, X., Meng, Z. & Cai, Y. (2011). *J. Hazard. Mater.*, *190*, 559.
[49] Xu, L. & Wang, J. (2012). *Environ. Sci. Technol.*, *46*, 10145.
[50] Hou, L., Zhang, Q., Jérôme, F., Duprez, D., Zhang, H. & Royer, S. (2014). *Appl. Catal., B.*, *144*, 739.
[51] Ling, Y., Long, M., Hu, P., Chen, Y. & Huang, J. (2014). *J. Hazard. Mater.*, *264*, 195.
[52] Lv, Q., Li, G., Sun, H., Kong, L., Lu, H. & Gao, X. (2014). *Microporous Mesoporous Mater.*, *186*, 7.
[53] Wang, W., Liu, Y., Li, T. & Zhou, M. (2014). *Chem. Eng. J.*, *242*, 1.
[54] Zubir, N. A., Yacou, C., Motuzas, J., Zhang, X. & Diniz da Costa, J. C., (2014). *Nature*, *4*, 4594.
[55] Wang, Y., Zhao, H., Li, M., Fan, J. & Zhao, G. (2014). *Appl. Catal., B.*, *147*, 534.
[56] Gu, L., Zhu, N., Guo, H., Huang, S., Lou, Z. & Yuan, H. (2013). *J. Hazard. Mater.*, *246–247*, 145.
[57] Huang, H., Lu, M. & Chen, J. (2001). *Water Res.* 35, 2291.
[58] Ortiz de la Plata, G., Alfano, O. M. & Cassano, A. E. (2010). *Appl. Catal., B.*, *95*, 1.
[59] Lu, M., Chen, J. & Huang, H. (2002). *Chemosphere*, *46*, 131.
[60] Hanna, K., Kone, T. & Medjahdi, G. (2008). *Catal. Commun.*, *9*, 955.
[61] Pouran, S. R., Abdul Raman, A. A., Wan Daud, W. M. A. (2014). *J. Clean. Prod.*, *64*, 24.
[62] Hanna, K., Kone, T. & Ruby, C. (2010). *Environ. Sci. Poll. Res.*, *17*, 124.
[63] Zhou, T., Li, Y., Ji, J., Wong, F., Lu, X. (2008). *Sep. Purif. Technol.*, *62*, 551.

[64] Matta, R., Hanna, K., Kone, T. & Chiron, S. (2008). *Chem. Eng. J.*, *144*, 453.

[65] Costa, R. C. C., Lelis, M. F. F., Oliveira, L. C. A., Fabris, J. D., Ardisson, J. D., Rios, R.R.V.A., Silva, C. N. & Lago, R. M. (2006). *J. Hazard. Mater.*, *129*, 171.

[66] Baldrian, P., Merhautová, V., Gabriel, J., Nerud, F., Stopka, P., Hrubý, M. & Beneš, M. J. (2006). *Appl. Catal., B.*, *66*, 258.

[67] Moctezuma, E., Zermeño, B., Zarazua, E., Torres-Martínez, L. M. & García, R. (2011). *Top. Catal.*, *54*, 496.

[68] Tao, T., Chen, Y., Zhou, D., Zhang, H., Liu, S., Amal, R., Sharma, N. & Glushenkov, A. M. (2013). *Chem. Eur. J.*, *19*, 1091.

[69] Munoz, M., de Pedro, Z. M., Casas, J. A. & Rodríguez, J. J. (2014). *Avances en Ciencias e Ingeniería*, *5*, 11.

[70] Munoz, M., de Pedro, Z. M., Casas, J. A. & Rodriguez, J. J. (2013). *Chem. Eng. J.*, *228*, 646.

[71] Dhakshinamoorthy, A., Navalon, S., Alvaro, M. & Garcia, H. (2012). *ChemSusChem.*, *5*, 46.

[72] Gao, L., Zhuang, J., Nie, L., Zhang, J., Zhang, Y., Gu, N., Wang, T., Feng, J., Yang, D., Perret, S. & Yan., X. (2007). *Nat. Nanotechnol.*, *2*, 577.

[73] Lu, A., Salabas, E., Schüth, F. (2007). *Angew. Chem. Int. Ed.*, *46*, 1222.

[74] Zhang, D. E., Zhang, X. J., Ni, Z. M., Song, J. & Zheng, H. (2007). *Crys. Growth Des.*, *7*, 2117.

[75] Wu, W., He, Q. & Jiang, C. (2008). *Nanoscale Res. Lett.*, *3*, 397.

In: Magnetite
Editor: Jacqueline Ward

ISBN: 978-1-63321-591-7

Chapter 3

CHARACTERISTICS OF MAGNETITE AND RELATED SPINEL FERRITE THIN FILMS, AND THEIR MAGNETORESISTIVE MULTILAYERS

Hiroshi Matsuda*
TOY BOX GALAXY, Kyoto, Japan

ABSTRACT

The oldest magnet, magnetite is the prototype material of spinel iron oxides which have been applied to an antenna, magnetic core, and magnetic recording medium. Since the magnetite film has not only the nature of the magnet, but electrical conductivity at room temperature, so thus the magnetite film having sub-micron to nanometer thickness deposited on the single crystal substrate, is expected as a useful material for the next-generation magnetic devices such as magnetic random access memory (MRAM).

In this chapter, our researches about magnetite thin films and its application are discussed. The high quality magnetite crystalline (epitaxial) films could be fabricated using a high-frequency magnetron sputtering method in the argon atmosphere with the oxygen-amount adjustment for the iron oxide target material.

* Corresponding author: e-mail: matsuda.hiroshi01@ms.naist.jp (eternal address given by NAIST)

Magnetite films show the very interesting physical properties; peculiar magnetic anisotropy (a nature that magnetization is easy to align along the crystal axis), and the metal-insulator transition (drastic changes in electrical resistance around 120K) depending upon their crystal properties. At first, interesting properties of the magnetite thin film are discussed. In the magnetic multilayer laminated in a few nanometers period composed of a non-magnetic and ferromagnetic material, a phenomenon whose electrical resistance changes drastically by an external magnetic field, so called "giant magnetoresistance (GMR)" effect is observed, and their multilayers are applied to the hard disk (HDD) read head having a high recording density over 100Gb/inch^2. In the magnetic multilayer using the magnetite as an electrode, the tunnel magnetoresistance (TMR) effect and large output (MR ratio over 10%) are observed at room temperature. Additionally, multilayers, called "spin valve", that are composed of magnetite as the magnetic pinning layer adjacent to the other ferromagnetic pinned layer such as iron, nickel, and cobalt, show large magnetoresistance (MR), that results in large output (several 10% MR ratio). In the second, the characteristics of magnetic multilayer using the magnetite thin film are discussed.

These multilayers using magnetite films also show very interesting magnetization processes (the external magnetic field dependence of the spontaneous magnetization), since the strong magnetic interaction between ferromagnetic metal and magnetite layer occurs in magnetoresistive multilayers, especially in spin valves. Finally, unusual magnetization reversals that could be seen in the magnetic multilayer using the magnetite thin film are also discussed.

1. Background

The oldest magnet, magnetite (Fe_3O_4) is the chief member of one of spinel iron oxides that have been applied to electronics using as an antenna, magnetic core, and magnetic recording medium. Fe_3O_4 (Lodestones) is known from BC, and has been used for the compass already in the era of the middle ages, also appeared in the book "*De Magnete*" written by William Gilbert in 17th century England [1].

In the 21st century, this oldest magnet has been studied for new electronic applications, is very interesting. Electronics using the "spin", which is the root of the magnetism, are called "spintronics". Magnetite is expected to be the useful material in spintronics, so many researchers make a great effort to study for practical applications.

The reason that magnetite is useful for the next generation spintronic material is as follows;

1. magnet having a spontaneous magnetization at room temperature
2. having electrical conductivity as metals at room temperature in spite of a metal oxide
3. being abundant in the crust having a large Clarke number

Magnetite is strictly called "ferrimagnetic" magnet, whose spins align in the crystal lattice so as to cancel each other such as "antiferromagnetic" materials, whereas spontaneous magnetizations have remained [2]. Many metal oxides are electrical insulators, but magnetite has electrical conductivity at room temperature due to having "free electrons" hopping in the crystal lattice.

These two features, "electric conduction" and "ferrimagnetism", are important characteristics in constructing a magnetic electronic devices. Magnets exhibit the phenomenon hysteresis, so memory properties appeared, for example, constitute a "1" and "0" of digital signal in the direction of NS poles of the magnet.

This property has been applied to classic magnetic recording tapes, now to high-density magnetic recording digital devices. Antiferromagnetic materials including ferrimagnet have a property called "pinning" to pin the direction of the spontaneous magnetization of nearest neighbor ferromagnetic materials in magnetic multilayer structures composed sub-micron to nanometer thickness thin materials [3]. This property leads to enhance "giant magnetoresistance (GMR)" effect, which is the electrical resistance change drastically with an external magnetic field in magnetic multilayers [3-4].

GMR has been applied to the read head of hard disk having high recording density in consumer electronics and personal computers. The electrical conductivity is important in driving the LSI. In magnetism, free electrons of ferromagnetic metals showing the electrical conductivity have the peculiar magnetic property called "spin polarization". Large GMR by "tunnel effect" of spin conduction through very thin insulator in magnetic tunnel junctions could be obtained using ferromagnetic materials having large spin polarization, results in realization of next spintronics devices such as magnetic random access memory (MRAM) [3]. This is the reason why magnetite is expected as the "spintronic" material.

We describe in detail about the basic properties of magnetite, and spinel iron oxides [2]. Fe_3O_4 has the same crystal structure as that of the spinel

($MgAl_2O_4$), cubic structure (crystal lattice constant a=8.39Å) at room temperature, and has peculiar arrangements of iron and oxygen ions. Iron ions are distinguished between A-site (Fe^{3+}) surrounded by four oxygen atoms and B-site surrounded by six oxygen atoms, while in the B-site Fe^{2+} and Fe^{3+} are randomly placed on composition ratio of 1:1.

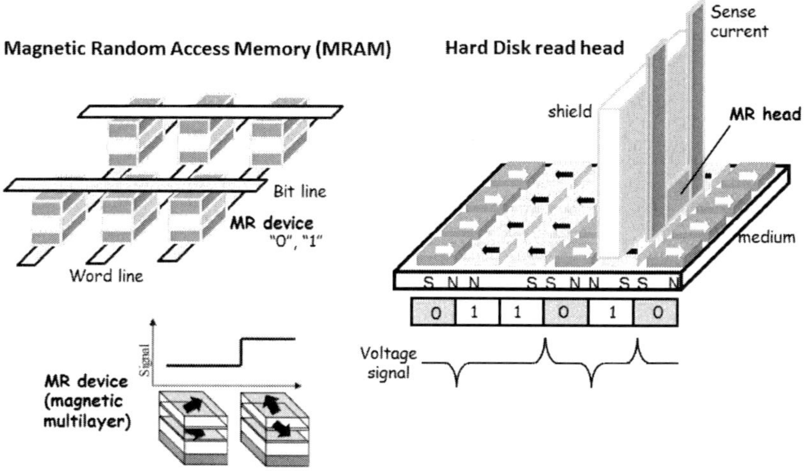

Figure 1. Magnetoresistive devices. (left) magnetic random access memory (MRAM) using tunnel magnetoresistive elements, (right) hard disk read head using spin valves.

According to Louis Néel [2, 5], strong magnetic coupling, called "super-exchange interaction" between A- and B-site through oxygen occur in the magnetite crystal, where cancelling the magnetic moments each other show "ferrimagnetism" as shown in Figure 2. As a result a magnetic moment has $4\mu_B$ (μ_B: Bohr magneton), saturation magnetization of 480 emu/cm^3.

Curie temperature (ferromagnetic to non-magnetic transition temperature) is very high (860K), so Fe_3O_4 has a spontaneous magnetization at room temperature. Since cancelling effects among spins each other are observed in the magnetic structure, magnetite and related spinel iron oxides have both the nature of ferromagnetic and antiferromagnetic properties. Thus, they are expected as GMR pinning layer, which pin the nearest neighbor ferromagnetic layer in magnetic "spin valve" magnetoresistive devices, leading to high MR ratio (large output signal of read heads). Fe_3O_4 has the electrical conductivity. This reason is that free electrons of B-site iron ions are easy to move between Fe^{2+} and Fe^{3+} ($4\times10^{-3}\Omega$cm @RT) as shown in Figure 3. In non-magnetic metals such as copper, aluminum, their conductive spins are canceled each other. On the other hand, spins of ferromagnetic metal as iron, cobalt are spin-polarized, so thus magnetotransport properties related between electricity and magnetism are seen [3]. Magnetite is predicted "half-metal" where the spin is polarized completely [6], so is expected for realization of very large MR ratio in spin dependent tunnel junctions [3,6-7].

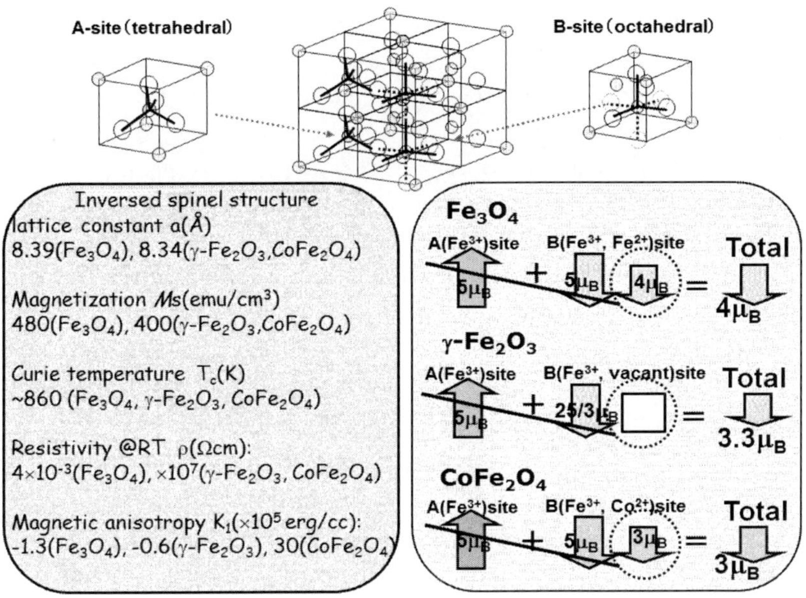

Figure 2. The crystal structure and their properties of magnetite, and related spinel ferrites.

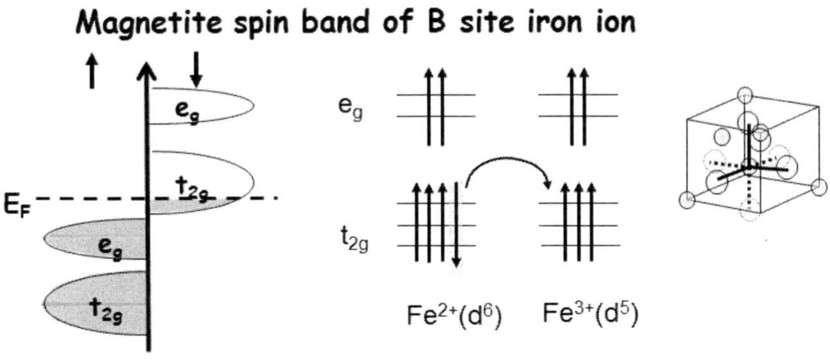

Figure 3. A schematic band structure of magnetite and its electronic structure. The spin electron of $Fe^{2+}(d^6)$ in B-site is able to move (hopping) to the orbital of $Fe^{3+}(d^5)$, So thus magnetite is predicted as "half-metal".

In contrast to the electrical conductivity of magnetite, maghemite (γ-Fe_2O_3) and cobalt ferrite ($CoFe_2O_4$) are electrical insulators. This reason is that electrons of Co (II) ions in the B-site in the case of $CoFe_2O_4$ cannot move in the lattice, and no free electron in the case of γ-Fe_2O_3 [2]. Maghemite has been applied to music recording tape in the form of fine particles. The nature called "magnetic anisotropy" (the property of magnetization tending to align along the crystal axis) is very strong in $CoFe_2O_4$ because large orbital magnetic moments of the Co ion are left, so thus $CoFe_2O_4$ is expected as an alternative material of rare earth-based permanent magnets [2].

Many of researchers including our group have been studied for application to new magnetic devices using magnetite thin films. Magnetite has served tunnel magnetoresistive devices using as the electrode and magnetoresistive spin valve devices using as the pinning layer. In this chapter, we discuss on characteristics of magnetite and related spinel iron oxide (ferrite) thin films, and their magnetoresistive multilayers.

2. Magnetite and Related Spinel Ferrite Thin Films

Many semiconductor materials to support modern electronics are applied with called "thin film" having the sub-micron to nanometer thickness deposited on a single crystal substrate. These elements are stacked over and over again, and microfabricated selectively by irradiating the ion beam, further are used as the tip of electronic circuits, finally applying to electronic devices such as the light emitting diode, laser, or dynamic random access memory (DRAM). Generally, thin films are formed through a process of evaporation particles such as atoms or ions to be deposited on the substrate plane, composed of a few microns or less thickness [8]. In the material formed as a thin film, specific characteristics different from the original bulk material appear such as quantum size effects. In particular, “perpendicular magnetization" films for high density magnetic recording medium have been attracting recently [9]. Thin films show a variety of characteristics.

Thin films are formed by particles adhering on surface of the substrate in vacuum. There are many methods for forming a variety of thin films. "Sputtering method" is particularly useful to a wide range from the fundamental study to industrial production [8]. Sputtering, particularly "high-frequency magnetron sputtering", is a method to deposition of target atoms on the substrates using the plasma discharge by applying a high frequency voltage between the chamber (ground) and the target material (cathode) as shown in Figure 4. Plasma discharge is generated on the target surface and concentrated at the target, whose components are deposited on the substrate (sputtering) in vacuum ($10 \sim 10^{-1}$Pa). The feature of this method is that the energy of the flying particles is very high (several eV (10^{4}K)), so thus it is possible to freely control the magnitude of deposition energy and fabricate homogeneous films at relatively low temperatures. Additionally, insulator films can be formed such as ceramic, fabrication of metal oxide crystal films having a large structural anisotropy, such as "epitaxial" films, are possible. Thin film characteristics are determined by the sputtering conditions, such as substrate material (crystal structure, surface cleanliness), substrate temperature and sputtering power (growth rate, energy), sputtering gas (composition, partial pressure ratio, and total pressure), target and electrode structure (target, substrate position). In this study, all thin films were made using a high-frequency magnetron sputtering apparatus (Advanced Technology Laboratories, Matsushita Electric Industrial Co. Ltd).

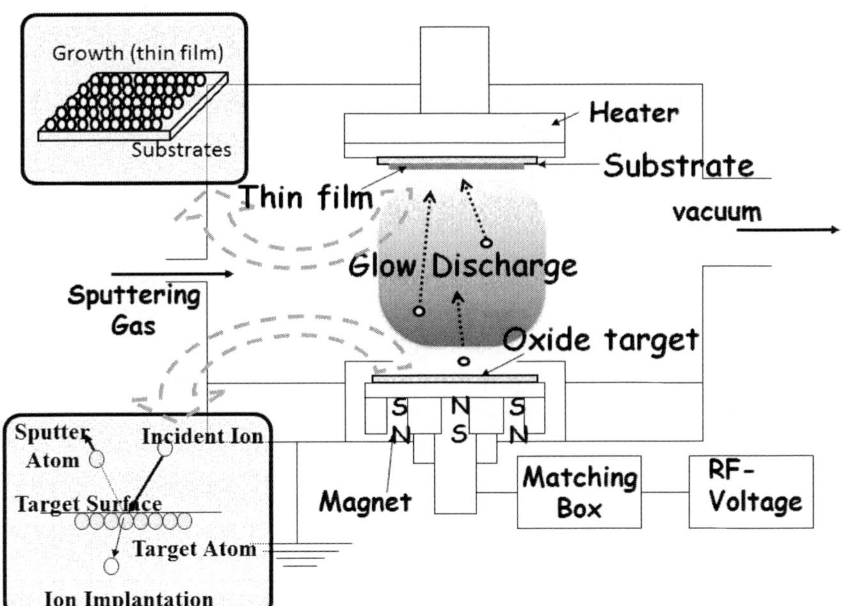

Figure 4. A schematic RF magnetron sputtering apparatus.

Magnetite films having various properties could be easily formed by sputtering. If the mismatch of lattice constant is small between the target material and substrate crystal structure, the single crystal (epitaxial) growth is possible. On a contrary, granular (polycrystalline) growth can also be possible using lattice mismatch. Magnetite films show various electric and magnetic properties depending upon their crystal properties, interesting properties of the thin film never seen in the bulk material. In this section, we focus on the magnetite and related spinel iron oxide thin films about their fabrications, basic and interesting properties such as uniaxial magnetic anisotropy, metal-insulator transition, biepitaxial growth, etc.

2.1. Magnetite (Fe_3O_4) Thin Films

It is possible to obtain a high-quality magnetite thin film nearly stoichiometric composition ratio using target a mixed powder iron and iron oxide (hematite) [10-13]. Variety of Fe_3O_4 films are obtained by adjusting the pressure in the argon atmosphere (~10^{-1}Pa), growth rate (3nm/min), substrate temperature (300°C). Observation the surface structure of Fe_3O_4 thin films (100nm thickness) with scanning electron microscope (JEOL JSM-7500F), the uniform surface no defect is seen in the case of magnetite deposited on the (110) magnesium oxide (MgO) single crystal substrate as shown in Figure 5(a). On the other hand, crystal grains having a grain size of 60nm are observed uniformly in the case of Fe_3O_4 deposited on surface oxidized silicon substrates, so thus magnetite becomes a granular film as shown in Figure 5(b). From the structural analysis by X-ray diffraction (Philips X'PERT), the Fe_3O_4 polycrystalline film on silicon substrate is confirmed by the observation of orientation multiple peaks because the oxidized silicon substrate has amorphous surface (in Figure 5(b)). On a contrary, it is found that the Fe_3O_4 grown on (110) MgO is the single crystal (epitaxial) film. The single crystal axis of (110) Fe_3O_4 is the same direction as one of the substrate with regard to in-plane orientation, results in the epitaxial layer with relation of <110>Fe_3O_4 //<110>MgO (in Figure 5(a)). Inverse spinel type cubic structure of Fe_3O_4 has the lattice constant of 8.39 Å, which is approximately twice that of NaCl type cubic MgO (a=4.21 Å) with a small mismatch of about -0.3%.

For the purpose of relieving the mismatching of crystal growth, insertions of second buffer layer called "seed layer" between the epitaxial layer and crystal substrate are performed well, thereon it is possible to obtain a "double heteroepitaxial" film can be formed [8, 10-11].

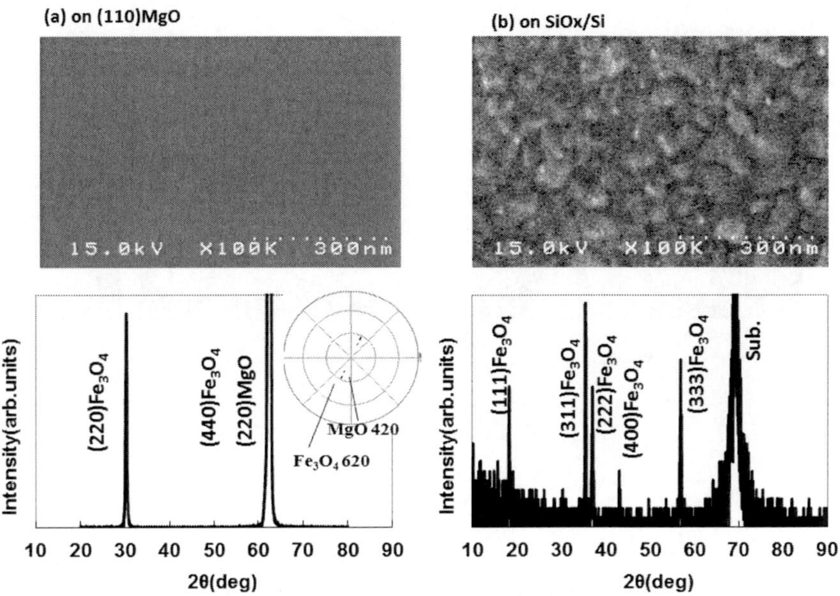

Figure 5. Surface morphologies and x-ray diffraction (XRD) patterns of magnetite thin films (100nm thick) deposited on (a) (110) MgO and (b) surface oxidized silicon substrates. The inset of XRD in (a) is the pole figure with regard to Fe_3O_4 620 and MgO 420 peaks.

In the epitaxial growth of magnetite, seed layers such as oxide and/or nitride are used. Double heteroepitaxial films of magnetite on MgO substrate are obtained as shown in Figure 6 using a cerium oxide (CeO_2) seed layer having a cubic fluorite structure (a=5.41 Å) [10] and titanium nitride (TiN) having a cubic NaCl structure (a=4.31 Å) [11].

The lattice constant of TiN is approximately 1/2 that of magnetite, although CeO_2 is not. However, this lattice constant is $1/\sqrt{2}$ times that of magnetite and $\sqrt{2}$ times of MgO, so thus only (110) magnetite is possible to growth on CeO_2 seed layer having a two-fold symmetric (110) plane on MgO substrate as shown in Figure 7(a).

Interestingly, the orientation of the cerium oxide seeded magnetite layer is perpendicular to the MgO substrate. Thus, the in-plane epitaxial relations are determined as <001>Fe_3O_4 //<110>MgO and <110>Fe_3O_4 //<001>MgO, whereas <110>Fe_3O_4 //<110>MgO and <001>Fe_3O_4 //<001>MgO for the no-seeded magnetite. The cerium oxide seed layer is found to rotate the upper magnetite lattice at 90 degree upon normal axis to the layer against the no-seeded magnetite as shown in Figure 6(b) [10].

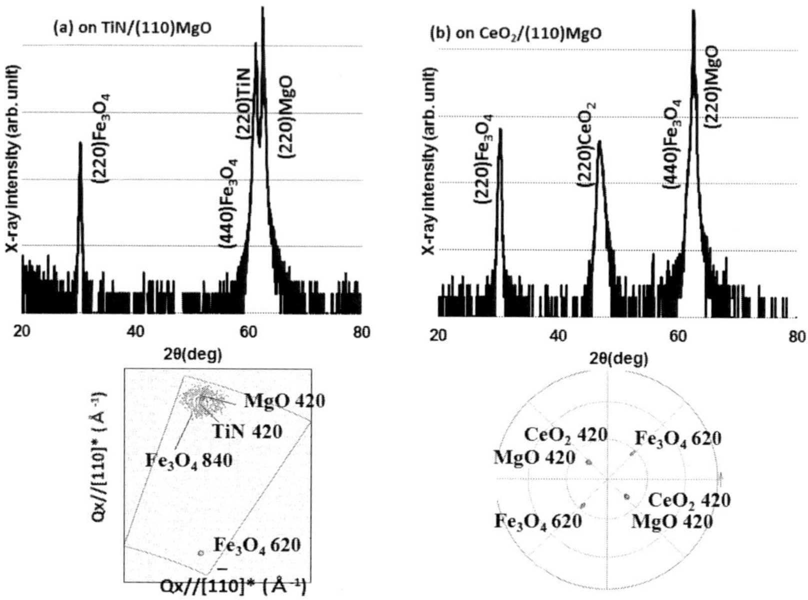

Figure 6. X-ray diffraction (XRD) patterns of double hetroepitaxial (a) TiN-seeded and (b) CeO_2-seeded Fe_3O_4 layers deposited on (110)MgO substrates. Additionally, the reciprocal mapping of TiN-seeded Fe_3O_4 and the pole figure of CeO_2-seeded Fe_3O_4 are also shown.

Homogeneous epitaxial layer of both CeO_2-seeded and no-seeded Fe_3O_4 layers can be seen in the cross-sectional images by transmission electron microscope (Topcon-EM002B), their electron beam diffraction patterns are corresponding to the crystal orientation of films as shown in Figure 8(a) and (c). Using these epitaxial growth properties, In-plane 90° rotated "biepitaxial" films are able to be prepared [10]. Biepitaxial grain boundaries are formed in the transition region between CeO_2-seeded and no-seeded Fe_3O_4 layers as shown in Figure 7(a). Oxide thin film devices have been studied having artificial grain boundaries as a quantum tunneling conduction barrier, for example superconducting Josephson junctions made of high temperature superconductive copper oxides [14], and tunneling magnetotransport junctions made of colossal magnetoresistive perovskite manganese oxides [15]. Therefore varieties of magnetite thin film devices are expected using the growth anisotropy of the oxide thin film, although large magnetoresistance could not obtained in our Fe_3O_4 biepitaxial grain boundary junction. Artificial grain boundary oxide films have been studied still now by minor researchers for the application.

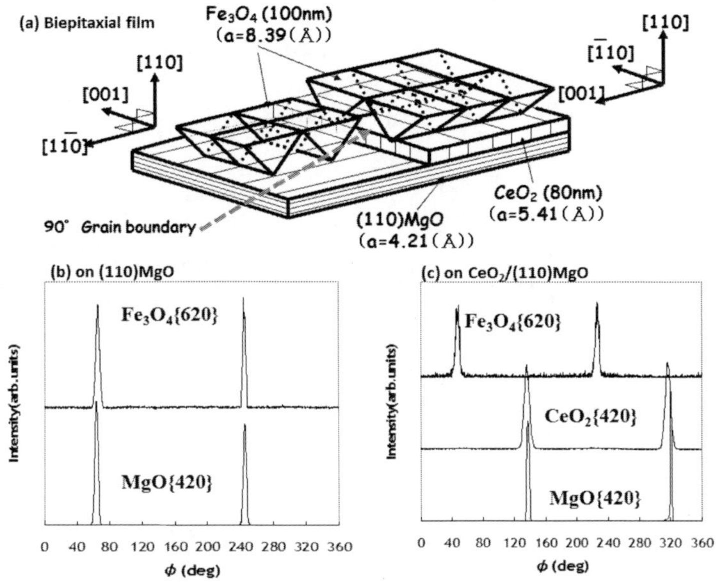

Figure 7. (a) A schematic illustration of the crystal lattice for 90° rotated biepitaxial Fe_3O_4 thin film on MgO (110) with partial CeO_2 seed layer. XRD φ–scans for (b) no-seeded Fe_3O_4 and (c) CeO_2-seeded Fe_3O_4 portions.

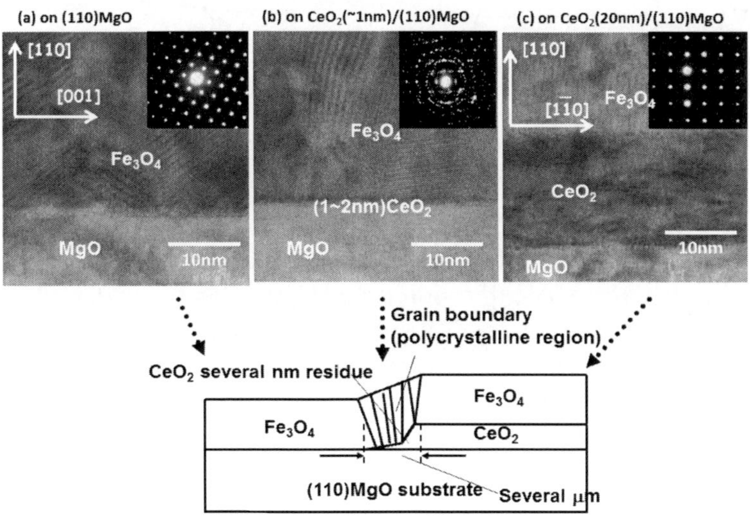

Figure 8. The transmission electron microscope (TEM) images for the different areas of the biepitaxial Fe_3O_4 film, (a) for no-seeded region, (b) for the boundary region, and (c) for the region with CeO_2 seed layer. The cross-section sketch of biepitaxial Fe_3O_4 films considered from TEM images.

Magnetic properties of the magnetite film are varying depending on the crystal structure. Magnetic hysteresis curves (external magnetic field dependence of the spontaneous magnetization) of the film were measured using a vibrating specimen magnetometer (VSM, Riken Denshi).

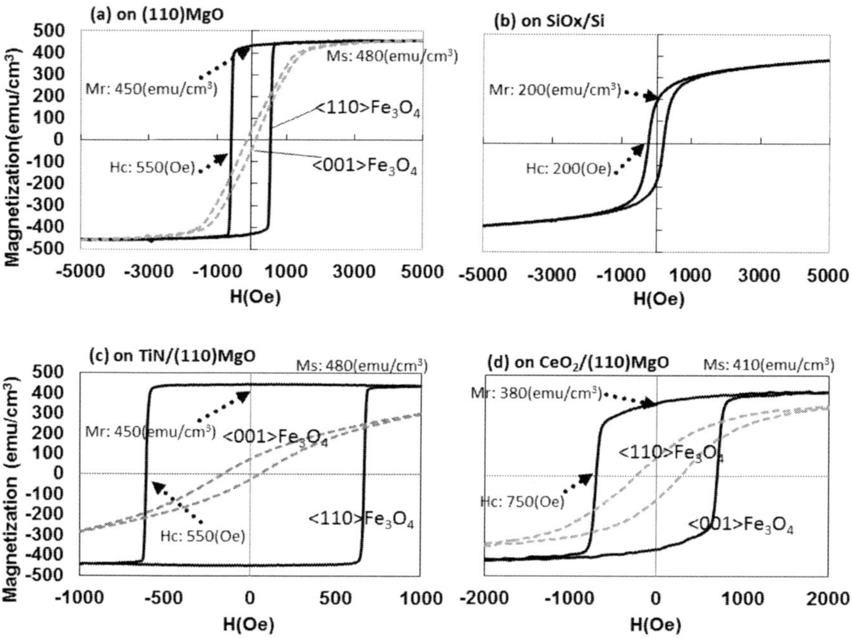

Figure 9. Magnetization hysteresis curves of Fe_3O_4 layers deposited on (a) (110)MgO, (b) surface oxidized silicon, (c) (110)TiN/(110)MgO, and (d) (110)CeO_2/(110)MgO substrate.

As shown in Figure 9, the dependence of magnetization hysteresis loop upon applied magnetic field direction is notably seen in the (110) Fe_3O_4 layer. Epitaxial Fe_3O_4 films show both magnetization curves of rectangular shape (black line) where large spontaneous magnetization Mr is remained at zero magnetic field and S-shape (gray line) where Mr is scarcely seen at zero field in the same sample as shown in Figure 9(a).

The coercive field Hc (the external magnetic field strength necessary to demagnetize the magnet) in the <110> direction is larger (550 oersted (Oe)) than that in the <001> axis (less 100 Oe). These characteristics are due to the magnetocrystalline anisotropy (a magnetic property tending to align easily along the crystal axis), especially markedly observed in two-fold symmetric (110) epitaxial Fe_3O_4 layer.

On the other hand, dependence on field applied direction are not observed in polycrystalline Fe_3O_4 layers deposited on silicon, so thus the magnetic anisotropy is not noticeable apparently as shown in figure 9(b). Additionally, Hc is very small (~200 Oe).

In (110)TiN-seeded and no-seeded magnetite films deposited on (110)MgO substrate, the magnetization easy axis indicating a square magnetization curve is in-plane <110> axis of Fe_3O_4 layers as shown in Figure 9(a) and (c), however the (110)CeO_2-seeded magnetite film has the easy axis along with <001> Fe_3O_4 layer as shown in Figure 9(d), are very interesting. In the biepitaxial Fe_3O_4 films, the steplike structure is observed in their hysteresis curves where the difference of coercive fields is 100-150 (Oe) for stoichiometric Fe_3O_4, 300-400 Oe for strontium-doped Fe_3O_4 as shown in Figure 10 [10].

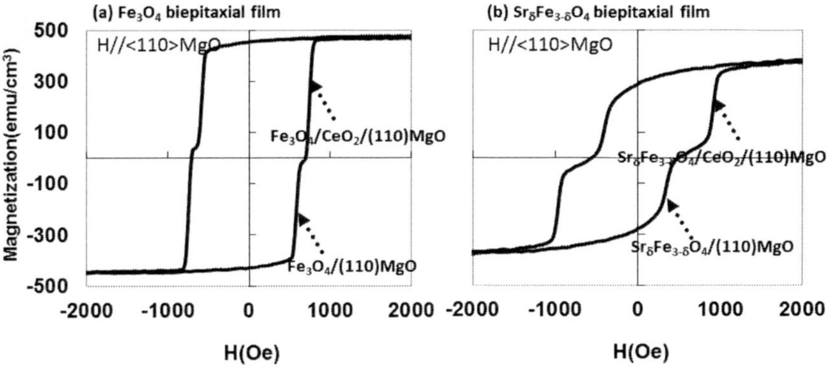

Figure 10. Magnetization hysteresis loops for the (a) biepitaxial Fe_3O_4 film with half a CeO_2 seed layer, (b) biepitaxial $Fe_{3-\delta}Sr_\delta O_4$ (δ=0.05) film measured at room temperature.

Magnetite layers show the dependence of the electrical resistance on magnetic field, i.e. magnetoresistance (MR) effect. MR measurements were performed using standard four-probe method in-plane direct current (10mA) with a gold electrode at 1mm intervals to sample pieces of 3 × 15mm square. As shown in Figure 11, the MR value of the polycrystalline Fe_3O_4 layer on the silicon is larger than that of (110) epitaxial Fe_3O_4 due to the inter-grain tunnel magnetoconductive effect [16]. Also, interesting properties could be seen about the temperature dependence of the MR, in which the peak of MR ratio is shown around 100K (in Figure 12 inset).

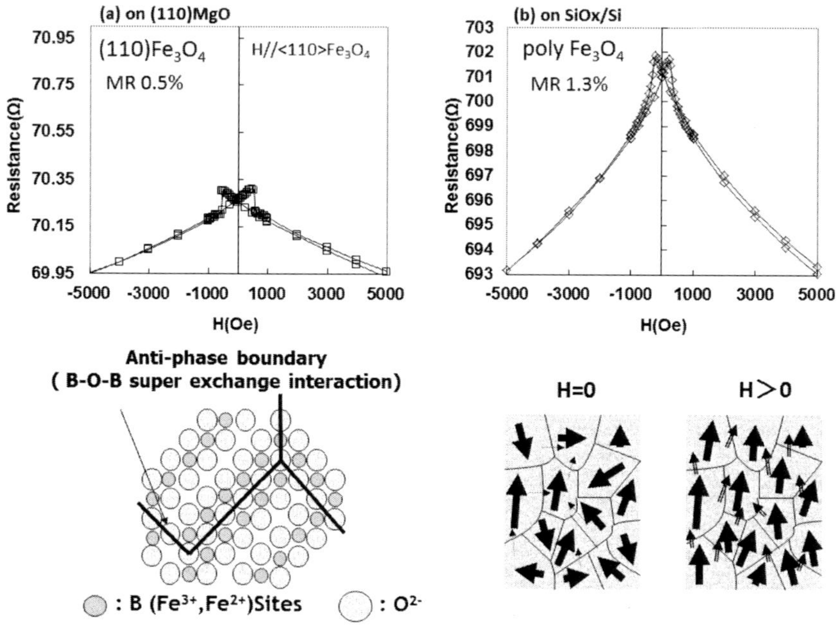

Figure 11. Magnetoresitance of Fe_3O_4 films deposited on (a) (110)MgO and (b) oxidized Si. Schematic illustrations of “Anti-phase boundary” of magnetite epitaxial films and inter-grain tunneling with regard to magnetotransport with and without applied external magnetic field are shown.

Magnetite is known to have the metal-insulator transition called “Verwey Transition” accompanied with drastic changing of crystal symmetry, magnetocrystalline anisotropy, specific heat, and elastic constant [2, 17]. Our Fe_3O_4 films also show Verwey transition as temperature dependences of electrical resistivity and coercive field of magnetization hysteresis loops [11]. Above mentioned MR jump might be due to this anomalous transition [16]. In generally, magnetoresistance are not seen in ferromagnetic metal single films such as iron, nickel, and cobalt, whereas our epitaxial Fe_3O_4 monolayers show MR ratio some extent due to “anti-phase boundary where transitions of iron and oxygen ions alignments in the magnetite crystal lattice are seen [7,18]. Recently, there are many studies for applied the magnetotransport sensor using these phase boundaries recently [7]. On the picture of the cross-sectional structure of Fe_3O_4 layer deposited on (110) MgO substrate (Figure 8(a)), it is confirmed that there is a transition in an oblique direction of 45 degrees with respect to the direction parallel and normal to the substrate.

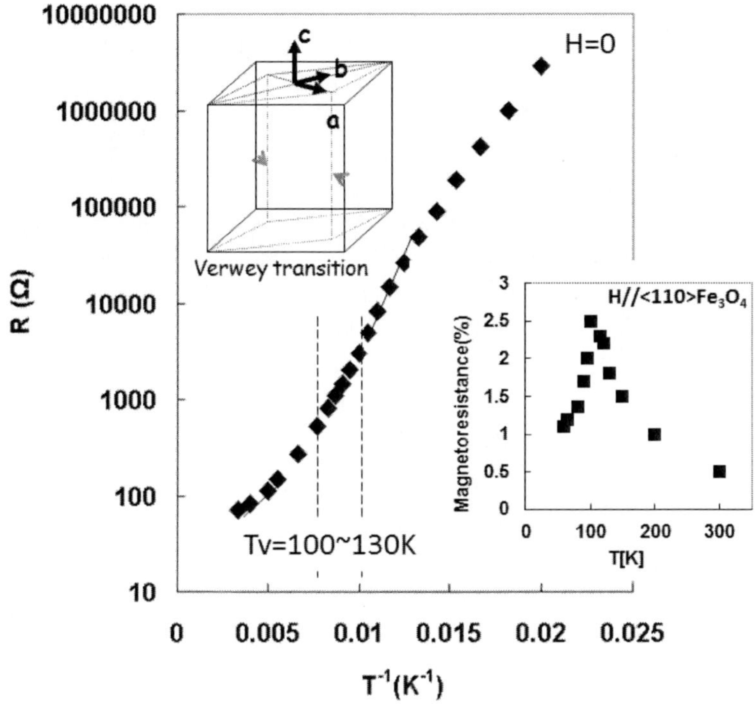

Figure 12. Temperature dependences of the resistance and magnetoresistance (inset) of the (110) epitaxial Fe_3O_4 film (100nm thick). A schematic illustration of Verwey transiton with regard to the magnetite crystal structure [2].

2.2. Other Spinel Iron Oxide Thin Films

Other spinel iron oxides except magnetite also have interesting properties. Upon heating to 400 °C under oxygen atmosphere, the black color Fe_3O_4 thin films on MgO single crystal substrate convert to the red color maghemite (γ-Fe_2O_3) films, and become the electrical insulator [12]. The Fe_3O_4 films having two-fold (110) and four-fold (100) symmetries on MgO substrate convert the ferromagnetic maghemaite layers, however the three-fold (111) Fe_3O_4 film on (111) MgO are oxidized changing from ferromagnetic to antiferromagnetic hematite (α-Fe_2O_3), result in disappearing spontaneous magnetization as shown in Figure 13. About maghemaite films, their coercive fields and saturation magnetization are diminished compared to magnetite.

Cobalt ferrite (CoFe2O4, CFO) films are also able to be prepared using target a mixed powder iron, iron oxide (hematite), and cobalt oxide (CoO) [19-20]. The lattice constant of inversed-spinel type cubic CFO (a=8.34 Å) is almost twice larger than that of NaCl-type cubic MgO (a=4.21 Å), which indicates good lattice matching with a small mismatch of 0.5%. Thus, the (110) CFO layer growths parallel to the surface of (110) MgO same as magnetite. The lattice constant of obtained CFO films from the x-ray θ/2θ-scan data is 8.33 Å and has slightly smaller than the bulk value with regard to out-of-plane, which indicates that the films are under compression normal to the plane and under slight in-plane tensile stress (in Figure 14(a)). Obtained CFO films are epitaxially grown on (110) MgO with the in-plane epitaxial relation of <110> MgO // <110> CFO and <001> MgO // <001> CFO. In-plane magnetization hysteresis of (110) CFO film shows both shapes of rectangular-shaped and S-shaped curves.

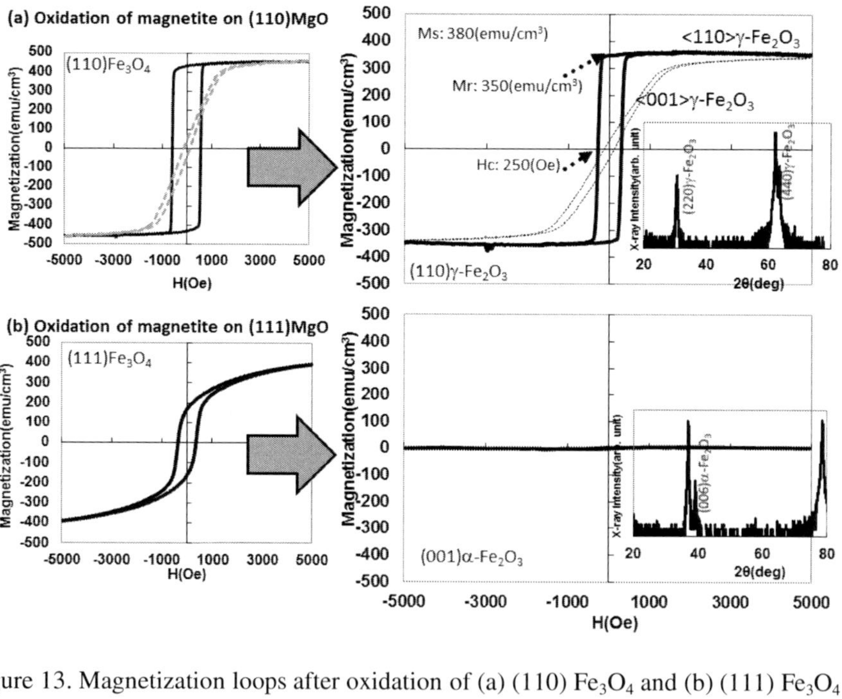

Figure 13. Magnetization loops after oxidation of (a) (110) Fe_3O_4 and (b) (111) Fe_3O_4 films.

The cubic magnetocrystalline anisotropy constant (K_1) of CFO is positive value [2], so the in-plane magnetization alignment along the <110> CFO direction should be a magnetic hard axis and the <001> should be an easy axis. However, rectangular-shaped loops with saturated magnetization of 400 emu/cm^3 are observed when the field is applied along the <110> CFO axis (black line), and linear S-shaped curve is also observed with the field along the <001> CFO axis (gray line), thus the present (110) CFO films have the in-plane easy axis parallel to <110> and hard axis along with <001> in spite of bulk K_1>0 of CFO as shown in Figure 14(b). The films could not be readily saturated in the hard direction and had considerably large coercive field values (~3500 Oe) in the easy direction, which indicates the presence of a large magnetic anisotropy in the CFO films. Our (110) CFO films might be under tensile stress in the (110) plane due to the lattice compression normal to the plain, so these effect of stress influence on the magnetic anisotropy of our CFO film. Since the saturation magnetostriction constant of a material CFO has a negative value ($\lambda_{100} \sim -590\times10^{-6}$) [2], the stress axis becomes a hard axis under tensile stress. This results in the (110) CFO film has a magnetic hard direction along the <001> axis and shows very different from bulk behavior, in other words, the uniaxial anisotropy term is a negative quantity.

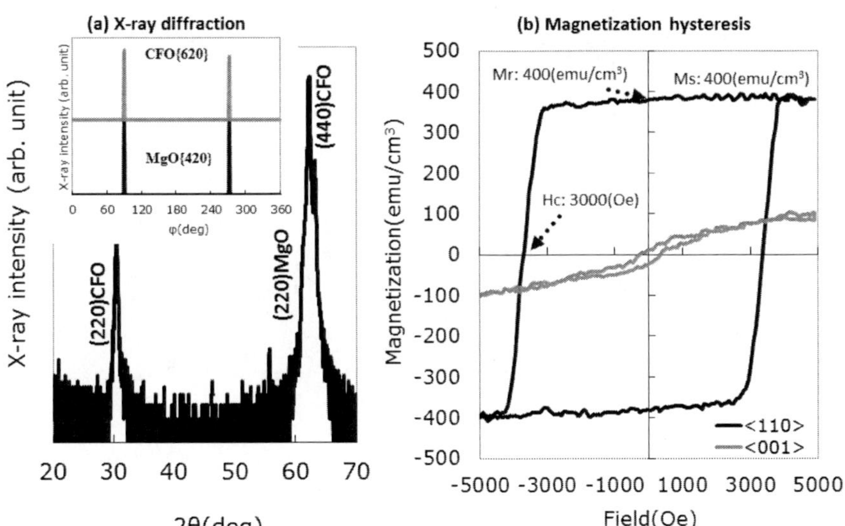

Figure 14. (a) The XRD pattern and (b) magnetization hysteresis loops of the (110) cobalt ferrite (CFO) thin film (80nm thick). Inset in (a) is XRD φ–scans for CFO 620 and MgO 420 peaks.

Interesting properties about spinel ferrite films could be seen, especially the (110) epitaxial spinel ferrite layers are also applied to the giant magnetoresistive multilayers as "tunnel junction" and "spin valve", that will be discussed in the following sections.

3. Large Magnetoresistance of Multilayers Using Magnetite Thin Films

Magnetoresistance (MR) effect, which is the electrical resistance is changed by applying external magnetic field, is observed in magnetic multilayer films composed of ferromagnetic and non-magnetic layers, in which observed MR ratio is very large. Definition of MR ratio as the changing resistance value divided by the resistance in applying high magnetic field, over 10% MR ratio are obtained at room temperature in magnetic multilayer films.

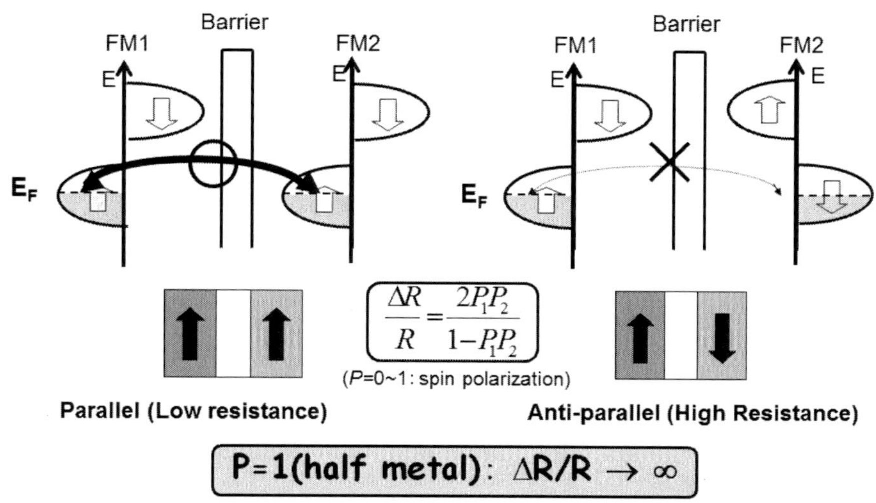

Figure 15. The Illustration of the magnetoresistance effect of the spin dependent tunneling transport. Two ferromagnetic half-metallic layers are separated by a non-magnetic insulator. There will be no current for the right case. For the left case there will only be an up spin current.

Magnetic tunnel junctions consist of two ferromagnetic electrodes (FM) separated by a non-magnetic insulator (I: Al_2O_3, MgO, $SrTiO_3$, etc.) as shown in Figure 15. In FM/I/FM junctions, the quantum tunneling current travelling from one FM electrode to the other could experience two distinct resistance stares depending on the relative magnetization configuration of the electrodes, resulting in the observation of very large MR ratio [3, 21]. There are two types of up (↑) and down spin (↓) of the conduction electrons, ferromagnetic properties of metals are determined by this ratio of the density of states with regard to up or down spins [3]. Non-magnetic metals such as copper, aluminum have canceled spins among each other in the Fermi level, however spins of ferromagnetic metal as iron, cobalt are spin-polarized, so thus their magnetotransport properties related between electricity and magnetism are shown [3]. Magnetite is predicted "half-metal" where the spin is polarized completely, so is expected for realization of very large MR ratio in spin dependent tunnel junctions [6-7, 22]. In the simplest case, the magnitude of the tunnel MR (TMR) is related to the spin polarization P of the individual ferromagnetic electrodes by Julliere model [21]:

$$TMR = \frac{\Delta R}{R} = \frac{2P_1P_2}{1-P_1P_2} \qquad (3\text{-}1)$$

where ΔR is the difference of resistance between parallel and anti-parallel magnetization configurations of two ferromagnetic electrodes, R is the resistance in the parallel configuration. Large TMR is expected by means of high spin polarized ferromagnetic electrodes for tunnel junctions. High spin polarized materials are known or have been predicted for several oxides, namely, half-metallic chromium oxide (CrO_2), mixed valence manganites such as (La,Sr)MnO_3, and magnetite (Fe_3O_4) (in Figure 16) [22]. Especially, magnetite is attractive material for its high Curie temperature compared to (La,Sr)MnO_3, so thus a better candidate for achieving large TMR at room temperature, that might realized next-generation spintronics devices such as high-density (over 100Gb/inch^2) magnetic recording reproducing head and non-volatile magnetic memory, such as a MRAM.

A phenomenon caused by spin-dependent scattering of conduction electrons at the interfaces between the ferromagnetic (FM) and the nonmagnetic (NM) layers in the magnetic layers, is known as giant magnetoresistsance (GMR) [3-4]. Therefore, the MR ratios depend upon the relation between the electron mean free path and the layer thickness of the magnetic and the nonmagnetic layers of the GMR films. Considering a three-layer structure of FM/NM/FM as shown in Figure 17, less scattering of the conduction electrons at the interface between NM and FM layers occurs when two ferromagnetic magnetizations are parallel. On a contrary, the antiparallel configuration increases the spin scattering. Thus, electrical resistance changes depending on the magnetization direction of two ferromagnetic layers, so GMR is obtained [3-4].

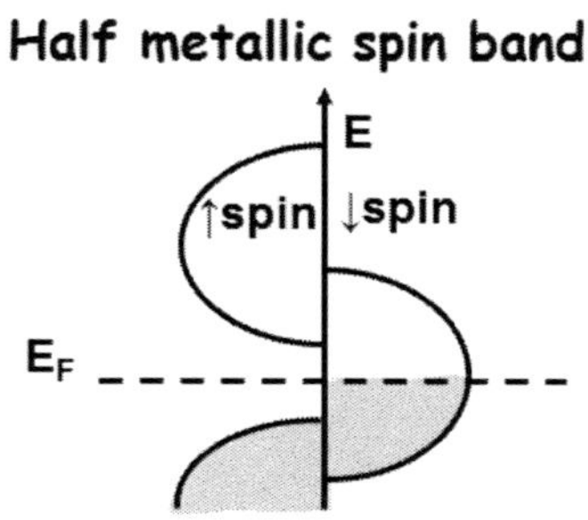

Spin polarization (P) and Curie Temperature (T_c) ** culc		
Materials	*P*(%)	T_c(K)
NiFe	37	713
Fe	46	1043
CrO_2	90	395
(La,Sr)MnO_3	95	360
Fe_3O_4	100**	853

Figure 16. A schematic band structure of half metals. Additionally, ferromagnetic materials of their spin polarization and Curie temperature are shown in the right table [22].

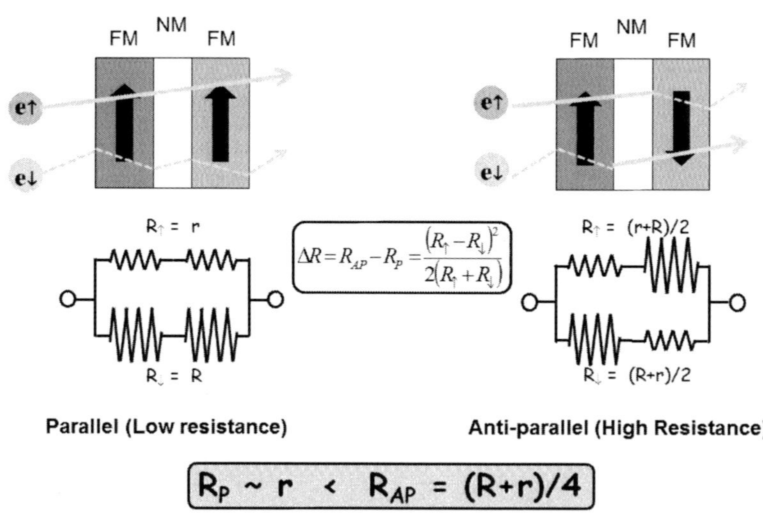

Figure 17. The Illustration of the giant magnetoresistance (GMR) effect with spin scattering. Parallel circuit models are represented as same physical system in upper illustrations. The magnetic layers are now represented by resistances R↑ and R↓. This shows very clearly that the total resistance for the two cases is different. In the left picture with parallel magnetizations the resistance for the spin up (spin down) electrons will be R↑ (R↓) in both magnetic layers. In the right picture with antiparallel magnetizations the spin up (spin down) electrons will have a resistance R↑ (R↓) [3].

Assuming that R↑ and R↓ are the up and down spin resistance experienced at a FM /NM interface respectively, the magnitude of the resistance change ΔR implying GMR is follows [3];

$$\Delta R = R_0 - R_H = \frac{(R_\uparrow - R_\downarrow)^2}{2(R_\uparrow + R_\downarrow)} \tag{3-2}$$

Thus, large GMR is obtained when the difference between R↑ and R↓ is larger. The spin-dependent scattering electrons in a three-layer structure can be represented by a parallel resistor circuit as shown in Figure 17. Large MR ratio as TMR are not to be obtained considering the theory, however GMR contribute to develop a great deal of modern storage technology. The discovery of GMR in Fe/Cr multilayer structures by the groups of Albert Fert and Peter Grünberg and co-workers, rewarded in 2007 by the Physics Nobel Prize [3].

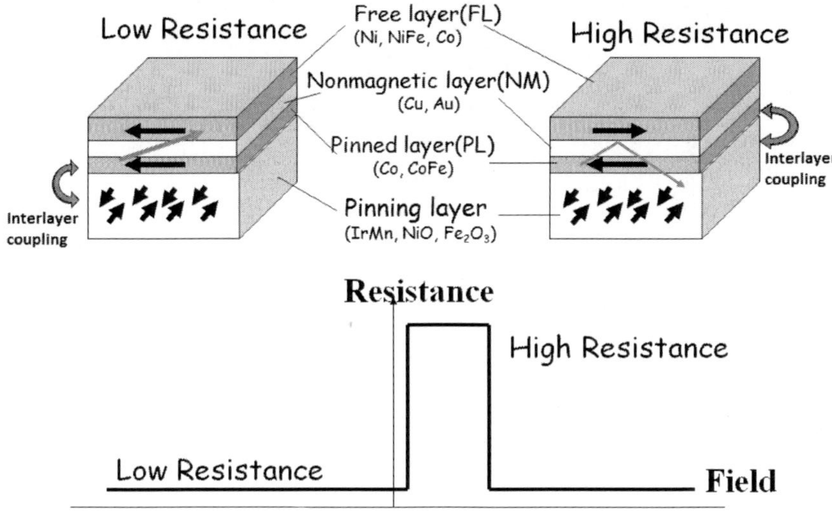

Figure 18. The illustration of the spin valve magnetoresistive elements. Pinning layer pins magnetizations of the nearest neighbor ferromagnetic pinned layer (PL). In the parallel state of both magnetizations, resistance is low. On a contrary, the antiparallel state results in high resistance.

GMR has been applied to a magnetic sensor, HDD of personal computer, etc. magnetic storage devices have been progressed through being higher recording density and enlargement of the information capacity. Drastic change of MR in low magnetic field is needed for increasing the output signal (S/N ratio) in the magnetoresistive element. Therefore, it is important to realize the fast switch between parallel and anti-parallel configurations to obtain a large MR ratio in low field [23]. Spinel iron oxide and magnetite has both the nature of ferromagnetic and antiferromagnetic properties. Thus, they are expected as GMR pinning layer, which pin the nearest neighbor ferromagnetic layer in “spin valve” magnetoresistive devices, leading to high MR ratio (large output signal of read heads) [11-12, 19-20]. Spin valves are basically composed of four layers, magnetic pinning layer / ferromagnetic pinned layer (PL)/ non-magnetic spacer layer (NM)/ ferromagnetic free layer (FL) as shown in Figure 18.

The exchange coupling occurs between the pinning and pinned layer, the pinning layer has the effect of pinning the magnetization direction of the ferromagnetic layer.

The magnetization direction of the pinned layer is fixed by the pinning layer, the large magnetic field is required relatively (over 100Oe) for the magnetization reversal. Accordingly, the exchange coupling between the pinning layer and PL is an important factor for determining the device performance, however the coupling mechanism is not still known enough about their spin alignments, learning to basic knowledge relationship of the data is useful.

Strong in-plane magnetic field direction dependences on the magnetization curve shape are seen in (110) epitaxial Fe_3O_4 films. The author presents in this section about the characteristics of the magnetoresistive multi-layers using (110) magnetite epitaxial layer applying to spin valves using as the pinning layer [12-13] and tunnel junctions using as the ferromagnetic electrode [11].

3.1. Spin Tunnel Junctions Using the Half Metallic Fe_3o_4 Layer

Magnetoresistance of spin-dependent tunnel junctions were studied using a (110) epitaxial Fe_3O_4 film [11]. Using multi-target sputtering apparatus (EIKO, Ltd.) in Figure 19, tunnel junctions are composed of (110) Fe_3O_4 (50nm)/ AlO_x (1.2nm)/ $Co_{75}Fe_{25}$ (10nm)/ Pt (30nm) on the TiN-seeded (110) MgO substrate.

In order to reduce the sheet resistance the conductive TiN (300nm) seed layer is deposited on (110) MgO single crystal substrate. After the aluminum layer deposition, AlO_x barrier layer is produced by subjecting a surface plasma oxidation.

Finally, the tunnel junction with a rectangular mesa of 4×12 μm^2 is fabricated using a photolithographic technique and Ar ion etching method. Magnetoresistance (MR) and field curves of the trilayer $Fe_3O_4/AlO_x/CoFe$ mesa are measured by physical properties measurement system (PPMS Model-6000, Quantum Design) using a standard dc four-probe method with the current (10μA) across the AlO_x tunnel barrier.

In the magnetoresistive characteristics of the tunnel junction, when the magnetic field applied to the in-plane easy magnetization axis direction (110) epitaxial Fe_3O_4 layer shows the largest MR over 10% in the coercive force difference between the CoFe and the Fe_3O_4 layer at room temperature as shown in Figure 20(b).

On a contrary, lower MR ratio is obtained along with hard axis direction as shown in Figure 20(a). It is important for enlargement of MR ratio that the parallel and anti-parallel magnetization state is realized, so thus it needs to enlarge the coercive force difference between the CoFe layer and the Fe_3O_4 layer.

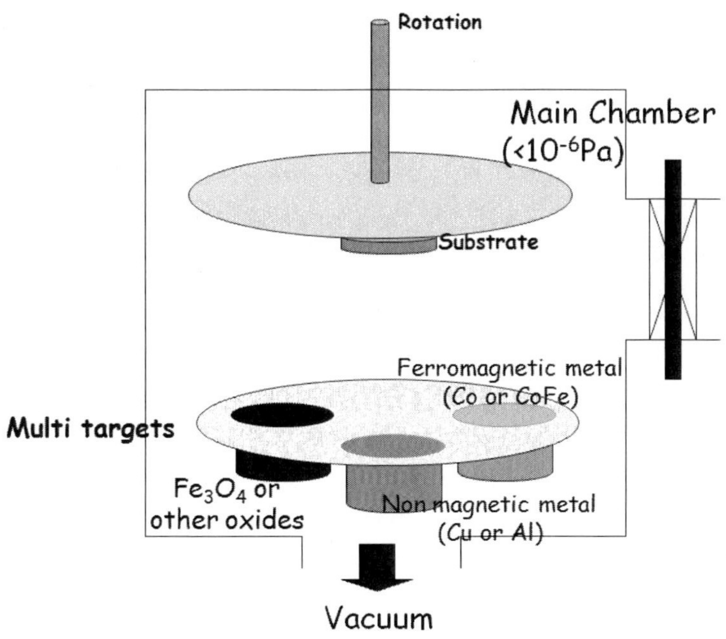

Figure 19. An illustration of multi-target RF magnetron sputtering apparatus.

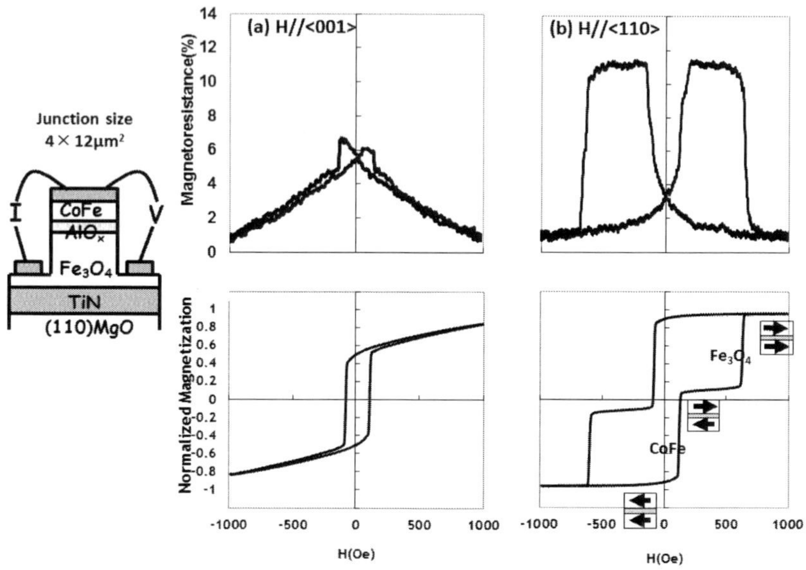

Figure 20. Magnetization and magnetoresistance hysteresis loops of the trilayer (Fe_3O_4/AlO_x/ CoFe) film and the spin dependent tunnel junction element. The magnetic field was applied along with in-plane (a) <001> and (b) <110> direction of the (110) Fe_3O_4 layer. An illustration of tunnel junction mesa composed of the trilayer. (Fe_3O_4/AlO_x/CoFe) film also shows.

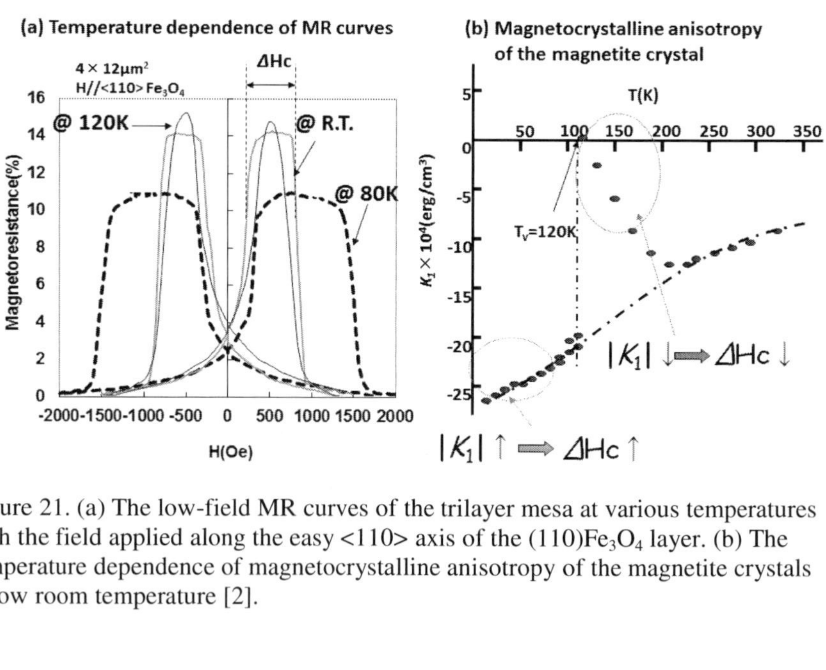

Figure 21. (a) The low-field MR curves of the trilayer mesa at various temperatures with the field applied along the easy <110> axis of the (110)Fe_3O_4 layer. (b) The temperature dependence of magnetocrystalline anisotropy of the magnetite crystals below room temperature [2].

Interesting phenomenon is also observed for the temperature characteristics of the TMR as shown in Figure 21(a). With decreasing the temperature the MR ratio remains almost constant down to nearly 100 K, while the hysteresis of the MR loop becomes narrower. Below 100 K, MR ratio decreases with the enlargement of the hysteresis width as the temperature decreases. These changes of magnetic and magnetoresistace properties might be caused by the Verwey transition due to drastic changes of the magnetic anisotropy and the elastic constants of Fe_3O_4 [2, 17]. The MR ratio represents the resistance difference ΔR divided the resistance R in the parallel state, i.e. ΔR/R. Comparing the temperature dependence of the resistance variation ΔR and R, the slope of the Arrhenius plot of the resistance is changed to the border near 100K, also the resistance variation ΔR show a gradual change in 100K as shown in Figure 22. The increment of the resistance R in zero magnetic field in the 60K ~ 100K (dlnR2) is ~ 2.5, however the increment of ΔR in the 60K ~ 100K (dlnΔR2) is ~1. So thus MR ratio represented by ΔR/R is decreased below 100K due to larger dlnR2 than dlnΔR2, whereas MR ratio does not change in higher temperature over 100K because dlnRΔ1 is the same value as dlnR1.

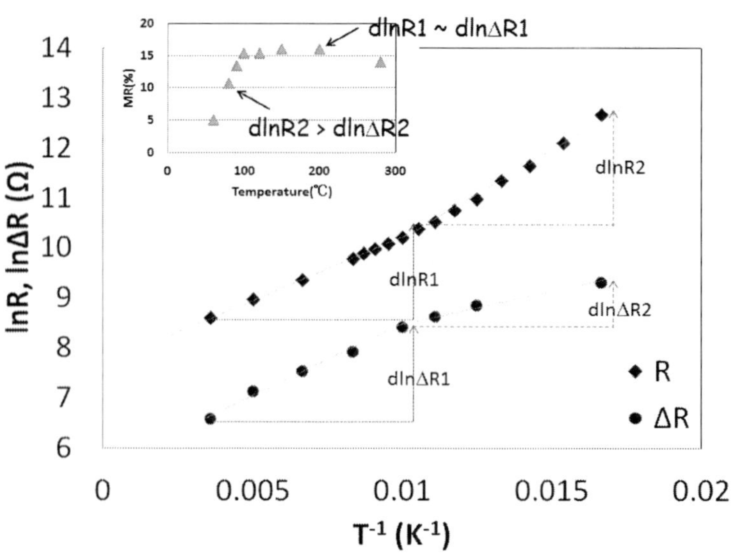

Figure 22. Temperature dependences of junction resistance R (at zero field), difference of resistance between parallel and anti-parallel states ΔR, and MR ratio (inset). From these Arrhenius plots, the MR diminishment of Fe_3O_4 tunnel junctions is confirmed below 100K.

In magnetite TMR junctions in this present study, large MR expected large spin polarization of Fe_3O_4 from (~ 100%) was scarcely observed. We believed the reason was due to the mismatch at the interface with the AlOx barrier and Fe_3O_4 layer. In this Fe_3O_4 tunnel junctions with alumina barrier layer, calculating the spin polarization P of Fe_3O_4 from formula (3-1) as the maximum MR ratio (18%) of this tunnel junction and spin polarization P~40% of $Co_{75}Fe_{25}$, spin polarization of Fe_3O_4 was about P~25%, was further different to expected value from the spin polarization of Fe_3O_4. Large MR ratio over 100% was obtained at room temperature in the epitaxial tunnel junction composed of Fe/ single crystal (001) MgO/ Fe, so feasibility of the MRAM is increased [3, 24].

By using an epitaxial MgO barrier layer, tunnel electrons called Δ1 band (k//=0) are selectively permeable moving in a direction perpendicular to the barrier, and large TMR could be obtained by conduction electrons (for Δ1 band) spin-polarized completely. A. Fert et al. had presented extremely large TMR ratio (1800%, @ 4.2K) in the junction composed of (La,Sr)MnO_3 /$SrTiO_3$ /(La,Sr)MnO_3 [25]. It is necessary to consider also the crystal lattice coincidence between the tunnel barrier and half-metal layer for larger MR ratio.

3.2. Spin Valves Using the Fe_3O_4 Pinning Layer

MR properties were studied about spin valves using magnetite and related spinel ferrites as the pinning layer [12-13]. Their multilayers show interesting properties. The sample are composed of the structure of (110) Fe_3O_4 pinning layer (50nm)/ Co pinned layer (PL, 2~10nm)/ Cu non-magnetic spacer layer (NM, 2.5nm)/ Co free layer (FL, 5nm) by multi-target sputtering apparatus (EIKO Ltd.). MR measurements are carried out by four-terminal method in-plane direct current (1mA) with a gold electrode at 1mm intervals in multilayer film specimens of 3 × 15mm square.

From the in-plane magnetization curves, it is confirmed that the epitaxial (110) Fe_3O_4 films have the easy axes along the <110> direction and the hard axes along the <001> direction as shown in Figure 9.

The spin valve using as pinning (110) Fe_3O_4 layer show anomalous MR curves as shown in Figure 23. Butterfly-shaped MR curves are observed applying field along the <001> direction, whereas two kind shapes of MR curves are observed in the same sample applying field along a <110> direction of Fe_3O_4.

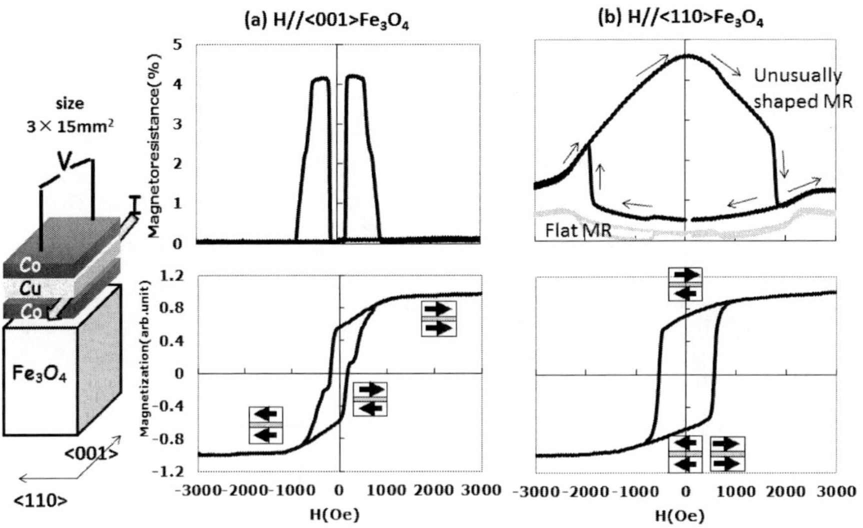

Figure 23. Mgnetoresistance and magnetization hysteresis loops of the magnetite spin valves composed of (110)Fe_3O_4(50nm)/PL(2nm)/NM(2.5nm)/FL(5nm). The magnetic field was applied along with in-plane (a) <001> and (b) <110> direction of the (110) Fe_3O_4 layer. An illustration of MR measurement of spin valves also shows. In the direction <110>, two types of the MR curve are seen in the same sample.

One is flat-shaped MR curves having low MR ratio, the other is the unusually (asymmetrically) shaped MR curves having high MR ratio being different from a flat-shaped MR curve.

About unusually shaped MR curves, which show large MR ratio in spite of the hard direction of the Co layer. This MR behavior might indicate the parallel and anti-parallel magnetization configuration between the top and bottom Co layers at zero fields.

In the spin valve sample of more than 5nm, only unusually shaped MR curve is observed, however flat shape MR curve (low MR) is not observed at all [13]. The observation of these two MR curves might be antithetical each other, so it should be considered that in-plane various magnetization reversals of PL and FL give rise to shape differences of MR curves.

For comparison, in the spin valve using the (110) γ-Fe_2O_3 pinning layer, it can be seen that strange characteristics related to the MR curve are similar to the case of Fe_3O_4 [12-13]. Maximum MR was about 4% in the Fe_3O_4 spin valve, whereas MR ratio of nearly 10% is obtained in the γ-Fe_2O_3 spin valve as shown in Figure 24. MR of the spin valve is dependent on the interface state of multilayers between NM and FM layer.

Pinning layers itself only pin the magnetization of the nearest neighbor ferromagnetic layer, not affect directly the MR ratio. Spin-dependent scattering might enlarge MR ratio using the insulating γ-Fe_2O_3 pinning layer as the specular spin valves using nano-oxide layers due to spin back scattering [26].

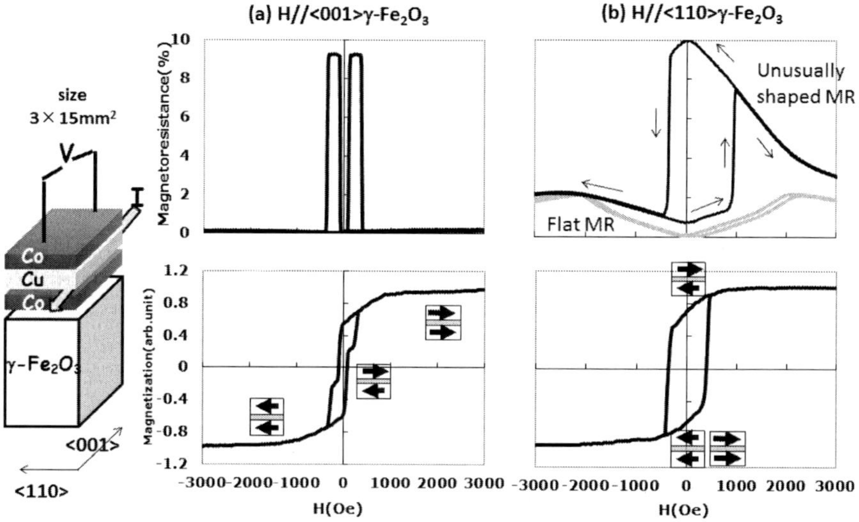

Figure 24. Mgnetoresistance and magnetization hysteresis loops of the magnetite spin valves composed of (110)γ-Fe_2O_3(50nm)/PL(2nm)/NM(2.5nm)/FL(5nm). The magnetic field was applied along with in-plane (a) <001> and (b) <110> direction of the (110) γ-Fe_2O_3 layer. An illustration of MR measurement of spin valves also shows. In the direction <110>, two types of the MR curve are also seen in the same sample.

It is important for enlargement of MR ratio in the GMR and TMR that the parallel and anti-parallel magnetization state is realized, so thus Fe_3O_4 and related spinel ferrites having large magnetocrystalline anisotropy are useful for realization of high performance magnetoresistive multilayers.

In the case of the spin valve, interlayer coupling between the pinning and pinned layers affects the MR properties. Called "90° coupling", where the magnetization of ferromagnetic layer is perpendicular to the spin alignments of pinning layer, might be considered in the present Fe_3O_4 and related spinel ferrite spin valves [12-13, 19-20]. We will discuss on the next section in detail.

4. Unusual Magnetization Reversals of Magnetite and Related Spinel Ferrite Spin Valves

Spin valves are required to have strong exchange coupling between the pinning and pinned layers for the control of their magnetization processes and the enlargement of MR ratio (in Figure 25(a)). Several pinning layers for spin valves have been studied such as antiferromagnetic iridium-manganese alloy (IrMn), hematite (α-Fe_2O_3), and orthoferrites ($RFeO_3$, R: rare-earth elements) [27]. Recently, multiferroic bismuth ferrite ($BiFeO_3$) has been studied for controlling the interlayer exchange coupling between the pinning $BiFeO_3$ and neighbor ferromagnetic layers by the electric field [28].

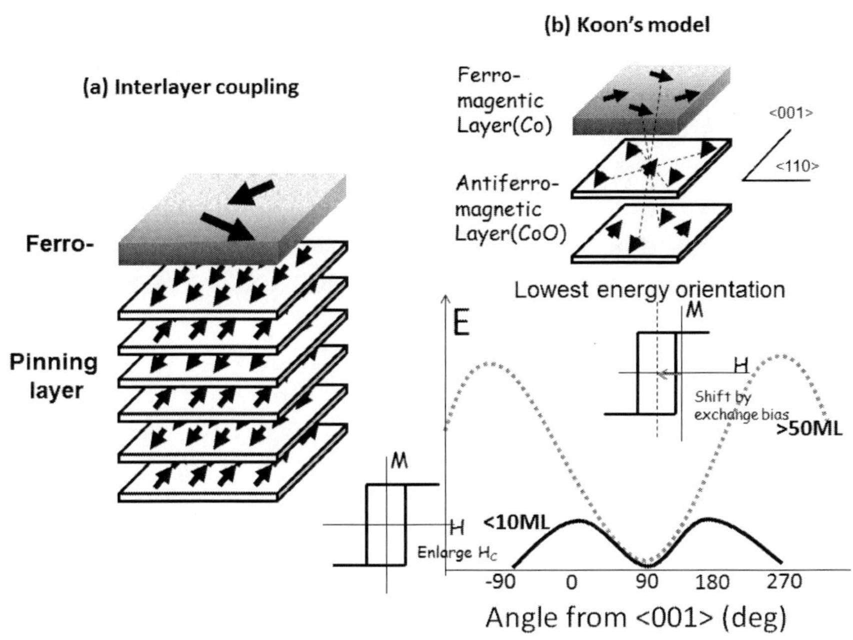

Figure 25. (a)The illustration of the bilayer composed of ferromagnetic pinned and pinning layer with interlayer coupling. (b)Koon's model about spin configuration near the interface plane for a Co/CoO bilayer with lowest energy orientation with perpendicular alignments, and energy as a function of angle between magnetization and <001> in the (110) CoO plane for Co/CoO bilayers with different the thickness of CoO [29].

There are many discussions for microscopic mechanisms of exchange coupling between the pinning and pinned layers. For example, "90° coupling model" has been discussed in many magnetic multi-layers. Koon showed in his study by mean field numerical calculations about magnetic multilayers that the most stable alignments of magnetizations between the ultrathin antiferromagnetic and ferromagnetic layers are oriented perpendicular to each other and give rise to the uniaxial anisotropy of the ferromagnetic layer [29] (in Figure 25 (b)). Van der Heijden *et al.*, also reported that the magnetic moments of the two Fe_3O_4 layers were aligned perpendicularly with respect to each other due to the 90° coupling between the bottom and top Fe_3O_4 layers over the ultrathin NiO spacer in Fe_3O_4/NiO/Fe_3O_4 trilayers [30]. Ijiri *et al.*, reported that the Co spins in exchange-biased CoO/Fe_3O_4 superlattices are frozen perpendicular to the Fe spins, this phenomenon was explained by the role of anisotropic exchange by Dzyaloshinsky-Moriya (DM) interaction between nearest neighbor tetrahedral Fe and octahedral Co at the interface of CoO/Fe_3O_4 [31]. In spite of the considerable interest in the exchange coupling in the pinning/pinned bilayers, its microscopic mechanisms are not fully understood.

Spinel ferrites as the ferrimagnet have also interested, so their spin valves as a pinning layer had been studied [12-13,19-20,32]. Especially, spin valves using $CoFe_2O_4$ performed as pinning layers had been studied and pinning mechanisms for spin valves using $CoFe_2O_4$ might be considered as strong exchange-spring like coupling between $CoFe_2O_4$ and pinned layers as SmCo/Fe bilayers [33-34].

However, $CoFe_2O_4$ pinning layers had polycrystalline and/or random orientations, details about the pinning mechanisms were not yet cleared. Since spinel ferrite epitaxial films are easy to fabricate by conventional film deposition technique, and thus are very suitable for studying properties of the oxide pinning layers for spin valves. As mentioned in the previous section, epitaxial (110) ferrite films as the pinning layer are showed a sharp rectangular magnetic hysteresis loop and uniaxial alignments of the in-plane magnetization, and the magnetization easy axis of Co layers might be perpendicular to that of (110) ferrite in the present spin valves. We discussed about magnetic processes of their spin valves with assuming 90° coupling between the pinned Co and pinning (110) Fe_3O_4 and related spinel ferrite layers and about the influence of the epitaxial ferrite layer on the neighbor Co layer.

It is interesting to study the influence of magnetization alignment of spinel ferrite pinning layers on that of pinned ferromagnetic layers. In this section, we discussed on interesting MR properties of the spin valves using epitaxial spinel ferrite as the pinning layer, especially the main concerns their interesting magnetization process.

4.1. MR Curves of Spin Valves Using a Cobalt Ferrite (CFO) Pinning Layer

In this section, magnetization reversals of the spin valve using the (110) cobalt ferrite (CFO) pinning layer are discussed. CFO spin valves show unusual properties similar to the magnetite pinning layer, although it is possible to explain easily using the numerical calculation by constructing a simple model about the complex magnetization reversals [19-20]. The spin valves were composed of (110) CFO (50nm)/Co (PL 2nm)/Cu (NM 2.5nm)/Co (FL 5nm) on (110) MgO substrate.

In-plane magnetization curves that the epitaxial (110) CFO films have the easy axes along the <110> direction and the hard axes along the <001> direction, whereas CFO spin valves show interesting MR properties (Figure 14). Butterfly-shaped MR curves, which implied the difference in the coercive fields between the top and bottom Co layers, are observed when the field is applied along the <001> direction of CFO as shown in Figure 26(a).

On the other hand, when the field is applied along the <110>, unusually (asymmetrically) shaped MR curves, which show large MR ratio in spite of the hard direction of the Co layer, are observed as shown in Figure 26(b).

This MR behavior might indicate the parallel and anti-parallel magnetization configuration between the top and bottom Co layers at zero fields. At a different point compared to the case of Fe_3O_4 spin valves, this asymmetrically shaped MR curve is observed regardless of PL thickness in CFO spin valves. A square hysteresis loop with small drops of magnetization at zero fields is observed along with <110>CFO, coercive fields are very large (4300 Oe).

Abrupt changes of MR ratio are seen at ±1000 Oe (MR switching field; H_{sw}) in the unusually shaped MR curves, however, no magnetization changes are seen at ±1000 Oe. So thus it is found that coercive fields in the magnetization loop are large values of 4300 Oe and not consist with H_{sw} (in Figure 26 (b)).

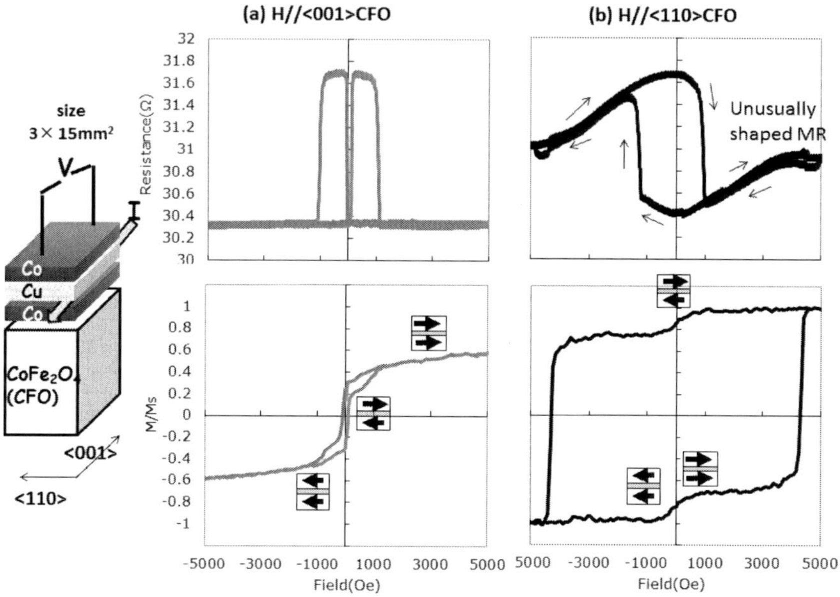

Figure 26. Mgnetoresistance and magnetization hysteresis loops of the magnetite spin valves composed of (110)CFO(50nm)/PL(2nm)/NM(2.5nm)/FL(5nm). The magnetic field was applied along with in-plane (a) <001> and (b) <110> direction of the (110) CFO layer. An illustration of MR measurement of spin valves also shows. In the direction <110>, only unusually (asymmetrically) shaped MR curve is seen.

Considering these results, next models of our (110) CFO spin valves could be assumed as follows as shown in Figure 27:

1. CFO pinning layers could be divided into two parts: one the thicker part (>40nm) named as "inner" apart from the bottom pinned Co layer (PL), the other the thinner part (<10 nm) named as "interfacial pinning region" nearest neighbor PL.
2. The thicker part "inner" has a large uniaxial magnetic anisotropy constant K_u (~3×10^6 erg/cm^3) and the thinner part "interfacial pinning region" has smaller K_u (~1×10^5 erg/cm^3), although saturation magnetization M_s of the "interfacial pinning region" is the same value as the "inner", 400 emu/cm^3.
3. CFO pinning layers have the continuous rotation of spins in the interface nearby the PL such as exchange spring magnets due to exchange stiffness coupling between "interfacial pinning region" having small K_u and "inner" having large K_u.

4. The interface spins of CFO (interfacial pinning region) might be perpendicular to the PL due to a presumably 90° coupling. The strong 90° interlayer coupling is assumed between PL and "interfacial pinning region" of CFO, while "inner" of CFO is not coupled perpendicularly with PL.
5. The ferromagnetic Néel-type coupling [35-36] between the top free Co(FL) and bottom pinned Co(PL) should be very weak, since both of parallel and anti-parallel configurations between FL and PL might be achieved at zero fields in unusually shaped MR curves with applying a field to the <110> CFO.

It should be mentioned about the Néel type interlayer coupling. In the interface between two ferromagnetic layers via a nonmagnetic layer, roughness may be inevitably formed in the multilayer films as shown in Figure 28(a). Ferromagnetic layers could weakly interact with another one due to magnetic pole appeared near roughness, leading to ferromagnetic coupling between two layers. This interlayer coupling is called Néel (orange peel) coupling. With the strong Néel coupling [35], the magnetization relative angle of the pinned and free layers does not change even if MR measurements [35-36], whereas in the present case Néel coupling might affect some extent.

Using numerical calculations called "micromagnetics" of the magnetization arrangement on the basis of above five models, it is possible to understand the complex magnetization process showing unusual MR curves.

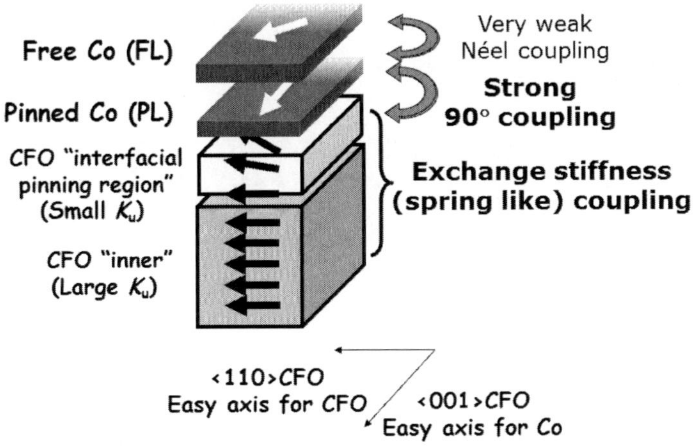

Figure 27. The model with regard to assumptions of magnetization processes of CFO spin valves for micromagnetic calculations.

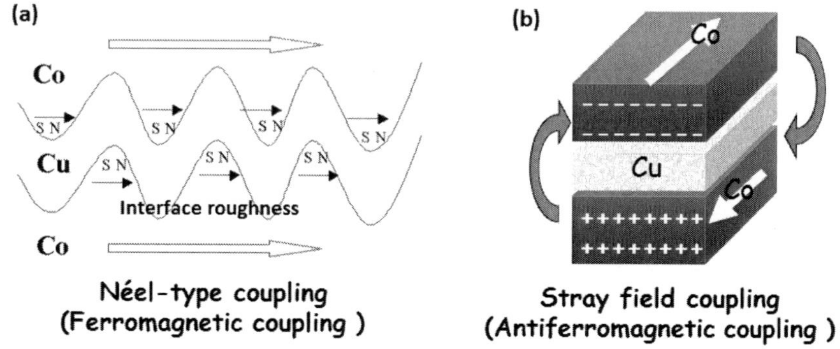

Figure 28. (a) Néel-type (orange peel) ferromagnetic coupling [35-36] and (b) stray field antiferromagnetic coupling between two ferromagnetic layers via a non-magnetic layer [37].

"Micromagnetics", named by W. F. Brown Jr [38], is a method of describing analytically the behavior of the magnetization of the ferromagnetic material having the finite shape, and calculating energetically stable configuration of the magnetization using a computer. Ferromagnets has some magnetic energies, external magnetic field (Zeeman energy), magnetic anisotropy, exchange interaction between the spins in the crystal lattice, and magnetostatic interaction by own magnetic dipoles.

Table 1. Material parameters using micromagnetic calculations of CFO spin valve's magnetization reversals. The magnetization alignments were determined with a static micromagnetic model that minimizes the total magnetic energy

	K_U(erg/cm^3)	Easy axis	M_S(emu/cm^3)	A(erg/cm)	thickness(nm)
Free Co(FL)	4.0×10^5	<001>CFO	1400	1.5×10^{-7}	5
Pinned Co(PL)	7.0×10^5	<001>CFO	1400	1.5×10^{-7}	2
CFO(interfacial)	1.0×10^5	<110>CFO	400	1.0×10^{-8}	10
CFO(inner)	3.0×10^6				40
Néel coupling constant: J_N = 0.01 erg/cm^2					
90° coupling constant: $J_\perp$ = 3.0 erg/cm^2					
<110>CFO →x direction, <001>CFO →y direction (long)					
X: 1μm, y: 5μm, Mesh size: 50×50×0.5nm^3					

Magnetization arrangements of ferromagnets take a most stable state by the competition of these magnetic energies. In the present case, interlayer coupling energies in multilayers are also added, for example, magnetostatic coupling between pinned and free ferromagnetic layers, 90 degree coupling between pinning and pinned layers. Numerical calculation about magnetization alignments of the spin valve applying the external magnetic field were performed using the Object Oriented Micromagnetic Framework (OOMMF 1.2alpha3 ver.) [39]. Material parameters using the calculation are shown in Table 1.

As shown in figures 29-32, good agreement with the experimental and calculated results of the MR curve and magnetization are obtained. The sketch of magnetization processes of CFO spin valves by micromagnetic calculations of is shown in Figure 30 and 32.

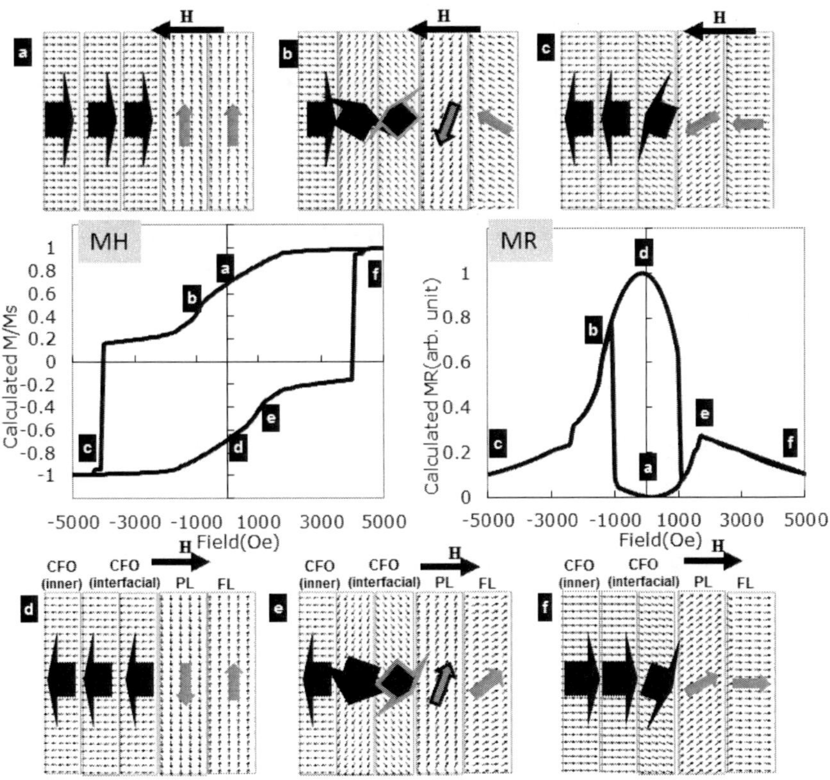

Figure 29. Calculated magnetization, MR curves, and magnetization reversals of the CFO spin valve for the field applied along <110> of CFO.

Considering the magnetization processes of unusually shaped MR curves with applying a field to the <110> CFO direction, it is plausible that the magnetization reversal of PL might occur at relative low H_{sw} due to strong 90° interlayer coupling between PL and "interfacial pinning region" of CFO having small K_u, resulting in abrupt changes of the MR ratio and the relative angle between PL and FL.

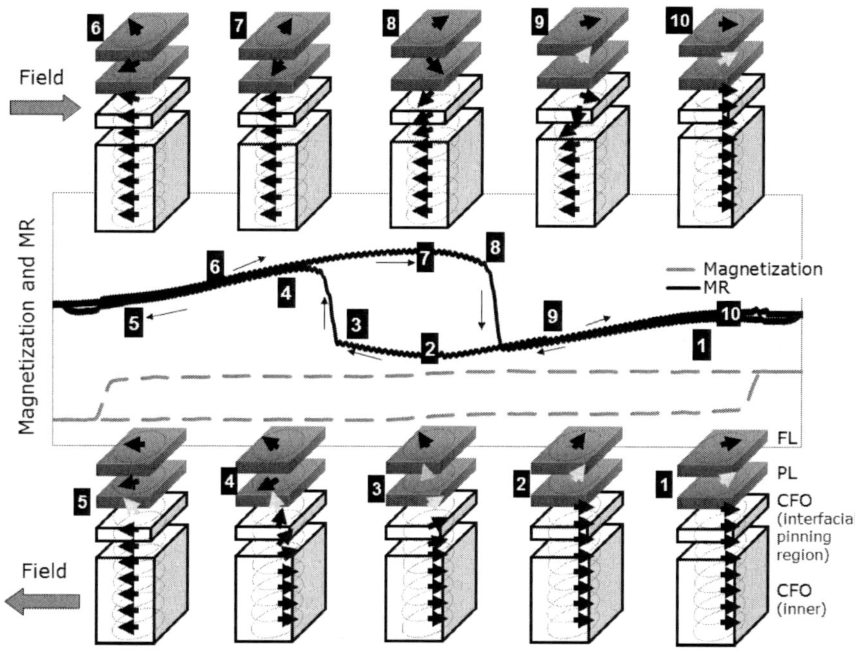

Figure 30. The sketch of magnetization reversals and experimental magnetization and MR curves of the CFO spin valve for the negative and positive field applied along <110> of CFO.

In this calculation, the thickness of the thinner part of CFO ("interfacial pinning region") is considered as 10nm. Since the thickness of "interfacial pinning region" is about 10nm, magnetization drops are seen in the calculated magnetization loop (magnetization stages 2 and 5 in Figure 29). Magnetizations of the "interfacial pinning region" exhibit the continuous rotation as a Bloch wall (the rotation of the spins in a plane parallel to the layer), however magnetization drops near 1000 Oe cannot observed in the experimental magnetization loop. In experiments, on a contrary, no change of magnetization is seen at all in spite of drastic MR changes as shown in Figure

30. If the "interfacial pinning region" of CFO is very thin (only few nano meters), these magnetization drops would not observed in experimental magnetization loops.

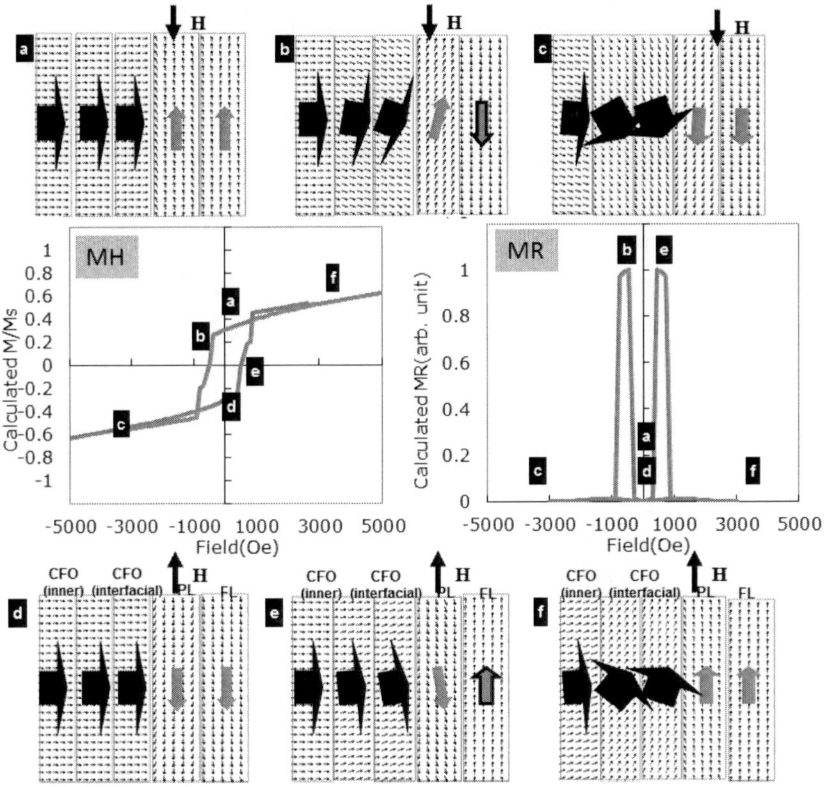

Figure 31. Calculated magnetization, MR curves, and magnetization reversals of the CFO spin valve for the field applied along <001> of CFO.

According to Koon, the ultra-thin antiferromagnetic (pinning) layer has been coupled perpendicularly with the neighbor ferromagnetic layer, which gives rise to the uniaxial anisotropy of the ferromagnetic layer [29]. There is a possibility where an antiferromagnetic interlayer (such as CoO and/or α-Fe_2O_3) lies between the pinned Co (2nm) and pinning CFO (50nm) layers in the present CFO spin valves. This might cause the MR switching at H_{sw}, which is different from the coercive field (H_c) of CFO having the large magnetic anisotropy constant. It seems that ultrathin (several atoms) intermediates play a role enough for the strong uniaxial anisotropy of the ferromagnetic pinned layer.

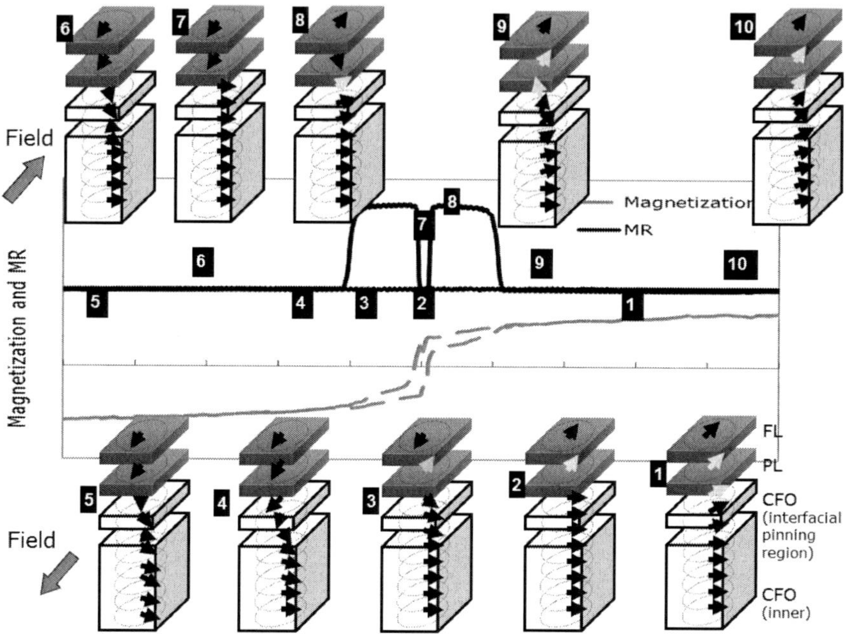

Figure 32. The sketch of magnetization reversals and experimental magnetization and MR curves of the CFO spin valve for the negative and positive field applied along <001> of CFO.

4.2. Unusual Magnetization Reversals of Spin Valves Using Fe_3O_4 Pinning Layer

In magnetite spin valve discussed on section 3-2, we presented interesting results about the strange phenomenon with regard to two kinds of MR curves observed in the same magnetic field applied in the <110> direction. In this and the CFO spin valves, cobalt FL and PL layers might have the magnetization easy axis along with <001> axis direction, whereas the hard axis of FL and PL is <110> axis as shown in Figure 23. Coercive field steps between PL and FL are observed in the magnetization curve of the spin valve in the <001> direction. On the other hand, MR values basically show the slight changing in both shapes of flat MR and unusually (asymmetrically) shaped MR applying field along with the <110> direction.

Considering these the magnetization reversals, it is easy to assume the magnetization process model that cobalt PL and FL might be perpendicular to the (110) Fe_3O_4 magnetization axis. Here, we discuss the details about this complex magnetization reversal of magnetite spin valve [13].

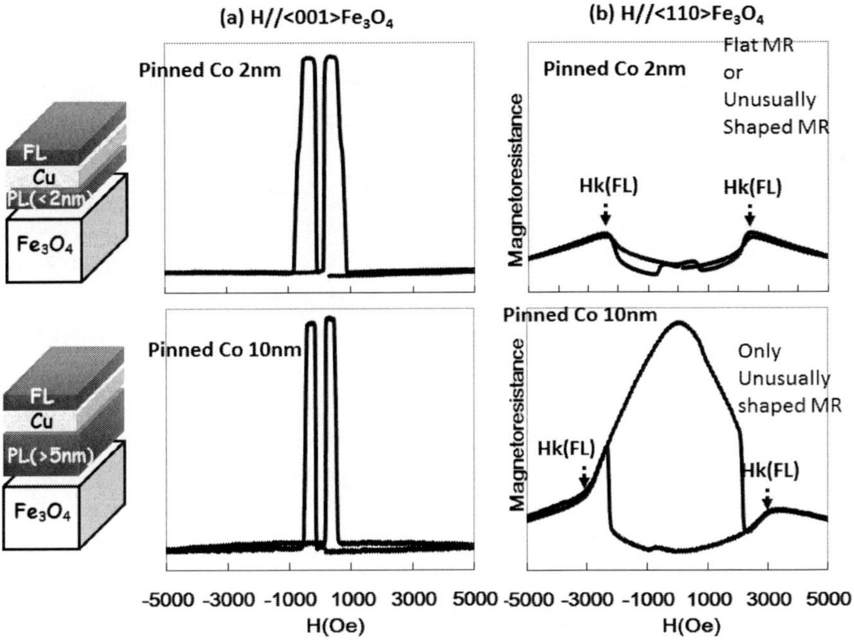

Figure 33. Mgnetoresistance hysteresis loops of the magnetite spin valves composed of (110)Fe_3O_4(50nm)/PL(t=2 and 10nm)/NM(2.5nm)/FL(5nm). The magnetic field was applied along with in-plane (a) <001> and (b) <110> direction of the (110) Fe_3O_4 layer. In the sample of PL2nm thickness, two types of the MR curve are seen in the same sample along with <110>. On a contrary, only unusually (asymmetrically) shaped MR curve is seen in the sample of PL5nm over thickness.

At first, MR curve of the spin valve are discussed changing PL thickness from 2nm to 10nm. As shown in Figure 33, in the sample of PL thickness of over 5nm, flat shaped MR curve of the <110> direction is not observed at all, whereas only unusually shaped MR is observed. It can be considered the same as CFO spin valve for this result. Considering the magnetization processes of this MR curve, the synchronous magnetization reversals (SR) of PL and Fe_3O_4 interlayer at switching fields occur due to 90° coupling between PL and Fe_3O_4 layers, resulting in the drastically change of MR ratio. On a contrary, drastic changes of MR ratio (relative angle between PL and FL) does not occur in the flat-shaped MR curve. This MR curve shows low MR ratio due to less change of relative angle between PL and FL during the magnetization process.

The observation of these two MR curves might be antithetical each other, so it should be considered that in-plane various magnetization reversals of PL and FL give rise to shape differences of MR curves. Since 90° interlayer coupling between Fe_3O_4 and PL might be affected regardless of the PL thickness, we considered that the difference of shapes of MR curves arises from the difference of magnetization processes of FL, and then the following two points were assumed about magnetization processes of two kinds of MR curves [13]:

1. About flat-shaped MR curves having low MR ratio, the relative angle between PL and FL would be less changing during magnetization processes, since synchronous magnetization reversals (SR) of FL might also occur as the magnetization of PL rotates synchronously, resulting in apparent less changing of the MR ratio as shown in Figure 35.
2. Unusually shaped MR curves having the large MR ratio and drastic MR changes at switching fields, might have both the parallel and anti-parallel magnetization configuration between PL and FL at zero field. After the SR of the magnetization of PL and the enlargement of relative angle between PL and FL whose SR do not occur, following the intensity of the field is changed to decrease down to zero, anti-parallel configurations of FL and PL are realized, resulting in the maximum MR ratio, and also the minimum MR vice versa as shown in Figure 37.

The ferromagnetic Néel coupling leads to be parallel between PL and FL magnetizations, whereas the antiparallel interaction called "stray field coupling" also occur between the two ferromagnetic layers via a nonmagnetic layer due to the magnetic poles of the sample edge as shown in Figure 28(b) [37]. In the present case, it should consider that stray field coupling occur between cobalt FL and PL as shown in Figure 34.

Here, magnetization processes of FL//PL/(110)Fe_3O_4 trilayer with regard to above assumptions (1) and (2) have been studied using the micromagnetic simulation [13]. It is important that the dependence of MR curves on the thickness of PL is confirmed under a fixed value of Néel coupling constant J_N. When thicknesses of PL are under 2nm, only contribution of Néel coupling between FL/PL would give rise to the magnetization process of the flat-shaped MR curve on the basis of an assumption of above (1).

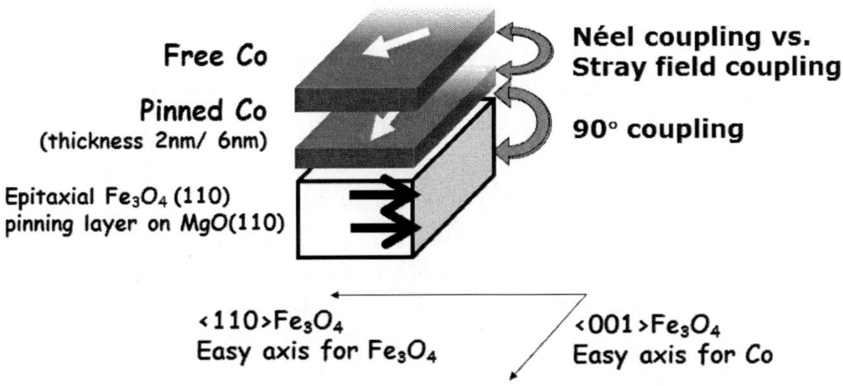

Figure 34. The model with regard to assumptions of magnetization processes of magnetite spin valves for micromagnetic calculations. In this assumption, the competition between Néel-type ferromagnetic and stray field antiferromagnetic couplings is considered.

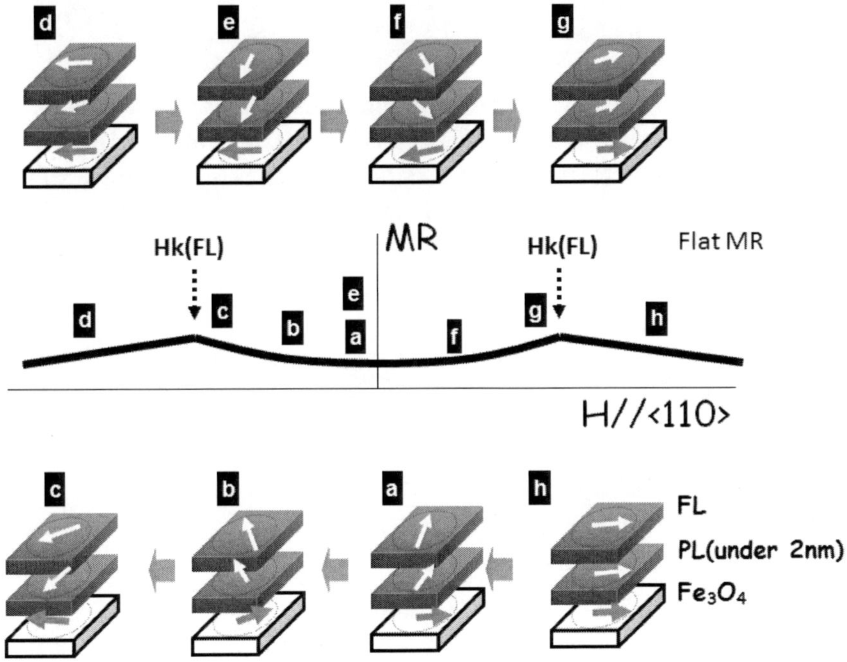

Figure 35. Schematic models of magnetization processes and the flat-shaped MR curve having low MR ratio. The relative angle between PL and FL is keeping small during magnetization processes, since synchronous magnetization reversals of FL might occur as the magnetization of PL rotates synchronously with PL/Fe_3O_4 interface, resulting in less changing of MR ratio.

On the other hand, in the case of (2), Néel (ferromagnetic) coupling due to interface roughness might compete with the stray field (antiferromagnetic) coupling due to free poles on sample edge in the thicker PL samples. Especially free pole at the sample edge of (thicker) PL might directly have an influence on the magnetization arraignment of FL, leading to the quasi-stable state in spite of the anti-parallel alignment at zero field.

Additionally, changing easily between parallel and anti-parallel conditions on the basis of an assumption of above (2), so thus the switch between low and high MR would be achieved during magnetization reversal, indicating that unusually shaped MR curves having high MR ratio would be seen over 5nm thickness of PL with applied field along to <110>. Especially, both parallel and anti-parallel conditions between PL/FL are stable, since the difference of total energies is zero. For the explanation of experimental MR curves, magnetization reversals were calculated with regard to thicknesses of PL 2nm and 6nm. We confirmed that the difference of total energies between parallel and anti-parallel conditions were zero in the fixed value of Néel (ferromagnetic) coupling constant.

Table 2. Material parameters using micromagnetic calculations of magnetite spin valve's magnetization reversals. The magnetization alignments were determined with a static micromagnetic model that minimizes the total magnetic energy

	K_U(erg/cm^3)	Easy axis	M_S(emu/cc)	A(erg/cm)	thickness(nm)
Free Co(FL)	5.0×10^5	<001>Fe_3O_4	1400	1.5×10^{-7}	4
Pinned Co(PL)	7.0×10^5	<001>Fe_3O_4	1400	1.5×10^{-7}	2 and 6
Pinning Fe_3O_4	1.0×10^5	<110>Fe_3O_4	480	1.0×10^{-8}	10
Néel coupling constant: J_N = 0.025 erg/cm^2					
90° coupling constant: $J_\perp$ = 3.0 erg/cm^2					
<110>Fe_3O_4→x direction, <001>Fe_3O_4→y direction (long)					
X: 1μm, y: 5μm, Mesh size: 50×50×0.5nm^3					

It is possible to discuss the complex magnetization reversals indicating this anomalous MR observation by micromagnetic calculations on the basis of above two models. Material parameters using the calculation are shown in table 2. As shown in figures 35-38, good agreement with the experimental and calculated results of the MR curve and magnetization are obtained.

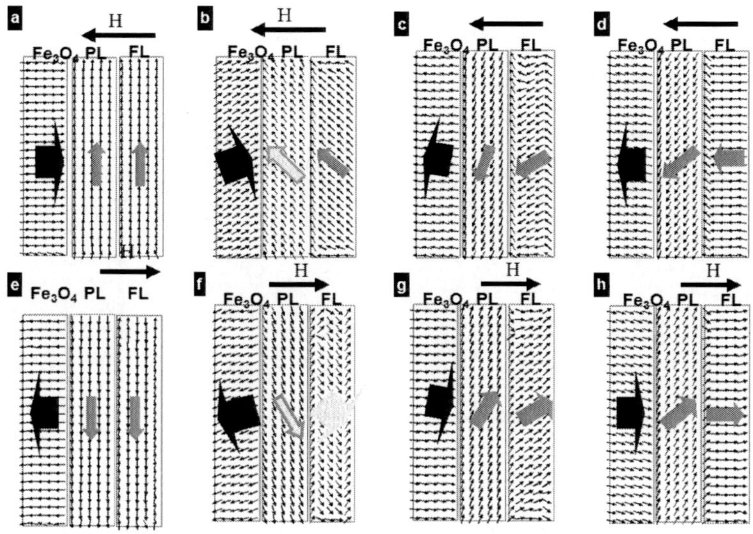

Figure 36. Calculated magnetization processes of the FL//PL(2nm)/Fe_3O_4 trilayer. Stages “a” to “h” are corresponded to Figure 35.

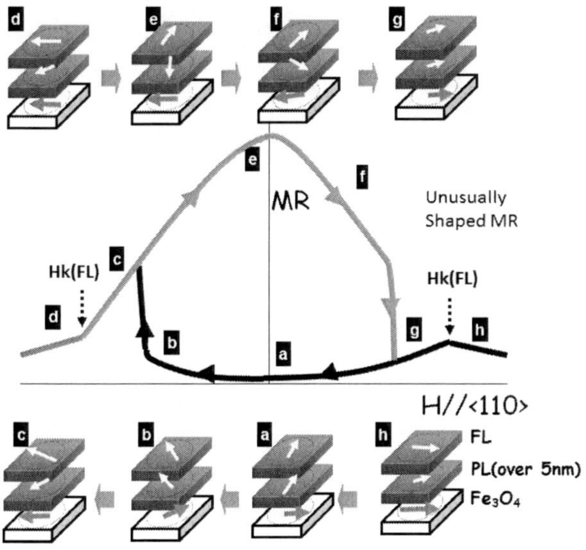

Figure 37. Schematic models of magnetization processes and the unusually (asymmetrically) shaped MR curve having high MR ratio. After SR of the magnetization of PL and the enlargement of relative angle between PL and FL magnetizations, the intensity of the field was changed to decrease down to zero, anti-parallel configurations of FL and PL were realized, resulting in the maximum MR ratio, and also the minimum MR vice versa.

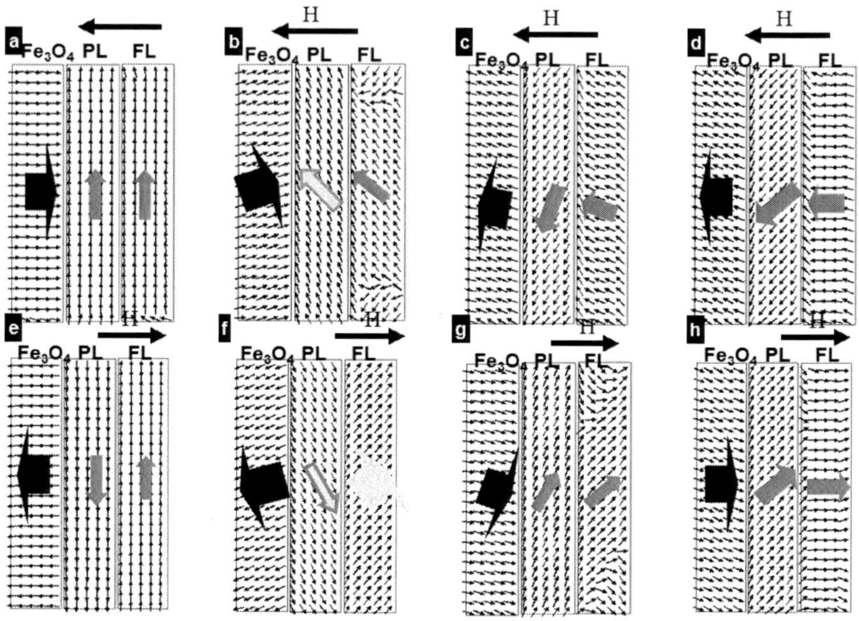

Figure 38. Calculated magnetization processes of the FL//PL(6nm)/Fe_3O_4 trilayer. Stages "a" to "h" are corresponded to Figure 37.

We can explain magnetization reversals for experimental observed MR curves with regard to thicknesses of PL as obtained results, so these discussions using micromagnetic calculations for the spin valve are plausible although considered sample size was 1×5 μm^2.

In addition, a very interesting phenomenon to exhibit two types MR curves in the same sample at PL2nm thickness is discussed. In the PL2nm thick spin valve, the unusually shaped MR curve was also observed depending upon the measurement procedure as shown in Figure 39.

First, the field was applied along the <110> direction just after the sample preparation, then the flat-shaped MR curve was observed. Next, the field was applied along the <001> direction resulting in the observation of butterfly-shaped MR curve.

Lastly, the field was applied again along the <110> direction, and the unusually shaped MR curve was observed as shown in Figure 39. The flat-shaped MR curve was never seen after this measurement. In this case, two different types of MR curves are observed by the measurement procedure in the sample in PL 2nm thickness applying field along the <110> direction.

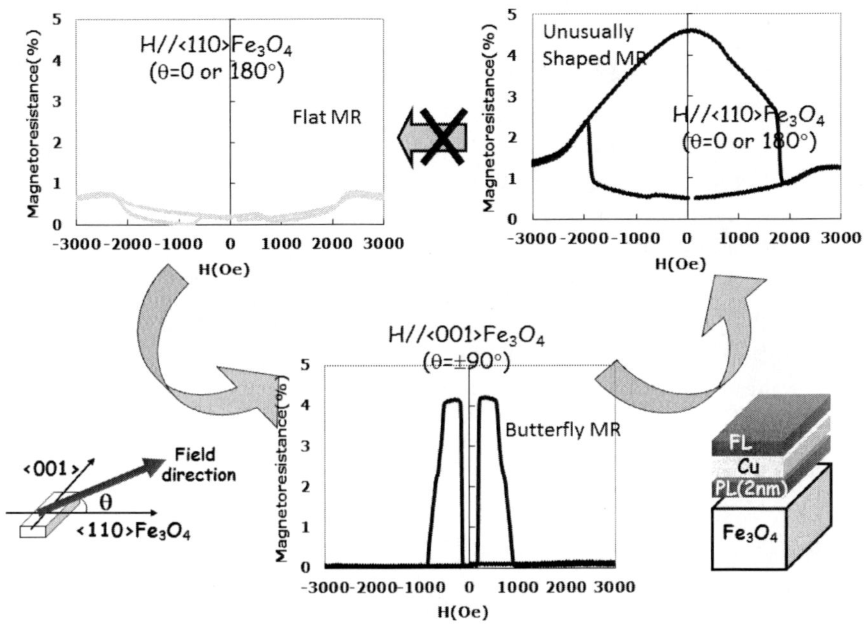

Figure 39. The measurement procedure about MR observation from flat-shaped to unusually (asymmetrically) shaped curves in the spin valve sample of PL2nm thickness.

It seems hard to discuss about this phenomenon, which might be explained by given as follows.

1. The magnetization configuration of the spin valve after preparation has a demagnetized state of in-plane multi-domain as shown in Figure 40. Applying field along with <110> direction, magnetization configuration of the spin valve would change to the highest energy "Multi-domain" state. Even applying external magnetic field <110> Fe_3O_4 direction after, the domain wall of PL and FL could not move due to cobalt layer having the hard axis. Therefore, synchronous magnetization reversals of the pinned layer with Fe_3O_4 pinning layer could not be carried out even if 90 degree coupling affect between the pinning and pinned layer. So thus, the flat-shaped MR curve would be observed. In this case, there is larger contribution of the domain wall energy more than 90 degrees coupling of PL/ Fe_3O_4.

2. Since the easy axis of cobalt PL and FL is along with the <001> Fe_3O_4 direction with observation the butterfly shaped MR curve, the in-plane magnetization reversal of cobalt pinned layer might be easily due to domain wall displacements applying along with the <001> axis [2]. After applying along <001> axis, the magnetization alignments "Intermediate" state of cobalt pinned and free layers, for example the single domain, would be shaped for reversible easily along <110> their hard axis as shown in Figure 40.
3. When the external field is applied along with <110> again, synchronous magnetization reversals (SR) of the pinned cobalt layer would be carried out easily together magnetization flip of Fe_3O_4 pinning layer at MR switching fields, due to larger contribution of 90 degree coupling than the wall energy. So thus, the asymmetrical shaped MR curve would be observed.

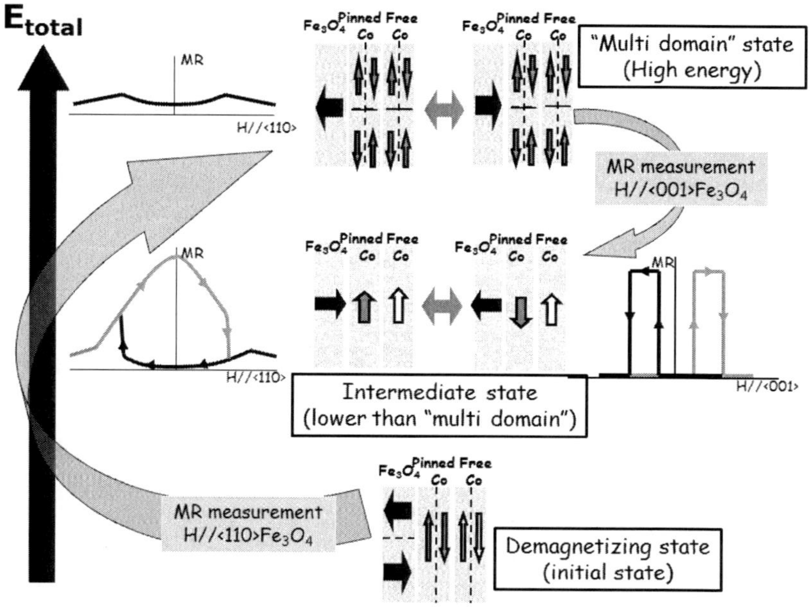

Figure 40. The illustration of the energy diagram about magnetization alignments of magnetite spin valves (PL2nm thickness) at zero field. The most stable alignment is "Demagnetizing (initial)" state. On the other hand, "Multi-domain" state related to the observation of the flat-shaped MR curve has higher energy than the "Intermediate" state related to the observation of the unusually shaped and butterfly shaped MR curves. So thus, shapes of observed MR curves might have the dependence on the measurement procedure.

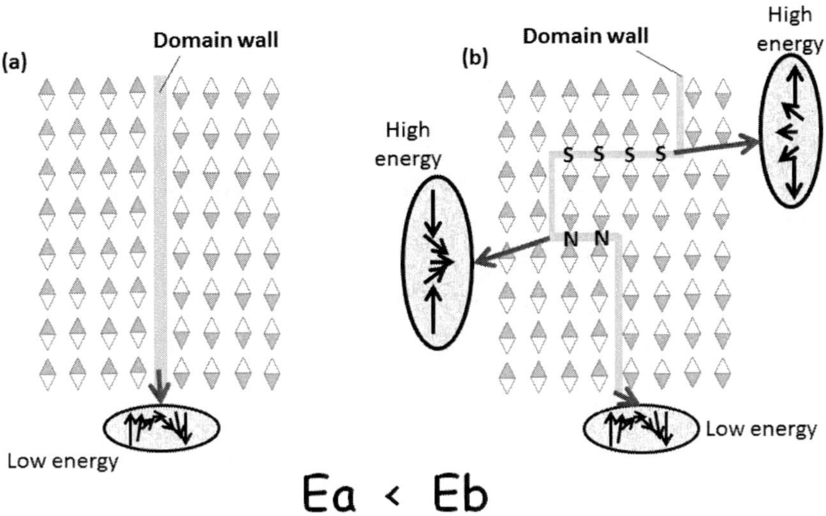

Figure 41. Néel domain wall structures in ferromagnetic thin films. The wall energy Ea (a) is lower than that Eb of (b) due to having opposite magnetic poles on domain walls as shown in (b) [2].

All ferromagnetic material have structures divided magnetization portions called "Magnetic domain", and the boundaries of them called "domain wall" are seen also due to decrease the magnetic energy [2]. Néel domain wall, such as to rotate spins in the plane is considered generally in magnetic thin films as shown in Figure 41(a). Sometimes, Néel domain wall structure would be forced to be an opposite pole pair and these free poles appear on the wall, giving rise to a demagnetizing field at domain walls having the higher energy as shown in Figure 41(b).

In this case as above the assumption (1), domain walls displacements might be difficult applying field along with <110> axis due to opposite poles [2]. In the spin valve having these multi-domain structures, the domain wall energy also cannot be ignored.

With regard to very interesting phenomenon observed in Fe_3O_4 spin valves, the competition with the interlayer coupling and domain wall energy takes place, so thus it might result in the observation of two shapes of MR curves in the same sample. However, considering the domain wall displacement with micromagnetic calculations, such a huge amount calculations are needed to match the experimental results and the discussion about these problems is difficult by the micromagnetic analysis.

Magnetization process can be seen in the spin valve using the spinel iron oxide, and it was found that magnetic multilayers using magnetite thin films show a very strange behavior by the author's studies. It may not be in such characteristics for practical, although it is significant that the knowledge about the magnetic coupling in the multilayer film which not yet been clarified was obtained in this study. In particular, interlayer 90° coupling is a common concept in exchange coupling between the ferromagnetic pinned and an iridium-manganese alloy pinning layer [40], which is used in the practical spin valve devices, we believe the great discoveries.

CONCLUSION

We have discussed mainly our research on the magnetoresistance effect in the multilayer using magnetite thin films. A variety of crystallinity of the magnetite thin film can be obtained by growth conditions, and other physical and magnetic properties are also very interesting. In the magnetic multilayer formed by laminating a few nanometers period and a non-magnetic and ferromagnetic layer, a large MR ratio is obtained at room temperature in the tunnel junction using magnetite electrode. Additionally, large MR ratio in spin valves using magnetite, maghemite, and cobalt ferrite as the pinning layer, which pin the nearest neighbor ferromagnetic magnetization, and is obtained also.

Anomalous properties with regard to MR jump and coercive field change in the magnetite tunnel junction appeared due to metal-insulator transition of magnetite in a low temperature, is very interesting. Magnetite and related spinel iron oxide as the pinning layer of spin valves, give the new knowledge about the interlayer coupling, 90 degrees between the pinning and pinned layers, so unusual magnetization reversals are observed. That is also interesting. Oxide magnetic materials including magnetite, and related spinel ferrite will expected for next electronic (spintronic) applications.

ACKNOWLEDGMENTS

These studies have been carried out in Nara Institute of Science and Technology and Matsushita Electric Industrial Co. Ltd (Panasonic, in present).

The author thanks Dr. Hideaki Adachi (Panasonic) and Dr. Sakakima Hiroshi (Ryukoku University) for useful discussions.

REFERENCES

[1] W. Gilbert, *De Magnete* (Dover Publications Inc.).

[2] S. Chikazumi, *Physics of ferromagnetism* 2nd edition (Oxford university press, New York, 1997).

[3] Fert, "The Nobel Prize in Physics 2007" Royal Swedish Academy of Sciences. 2007-10-09.

[4] M. N. Baibich, J. M. Broto, A. Fert, F. Nguyen Van Dau, F. Petroff, P. Etienne, G. Creuzet, A. Friederich, and J. Chazelas, *Phys. Rev. Lett.*, 61, 2472 (1988).

[5] L. Néel, *Ann. Physiq*. 3, 137 (1948).

[6] Yanase, L. Siratori, *J. Phys. Soc. Jpn.*, 53, 312 (1984).

[7] J. Moussy, *J. Phys.* D*: Appl. Phys.*, 46, 143001 (2013).

[8] K. Wasa, S. Hayakawa, *Handbook of sputter deposition technology* (Noyes publications, Park Ridge, New Jersey, 1992).

[9] Y. Murayama, *J. Phys. Soc. Jpn.*, 21, 2253 (1966).

[10] H, Matsuda, H. Adachi, A. Odagawa, K. Setsune, and H. Sakakima, *J. Mater. Res.*, 17, 1985 (2002).

[11] H. Matsuda, M. Takeuchi, H. Adachi, M. Hiramoto, N. Matsukawa, A. Odagawa, K. Setsune, and H. Sakakima, *Jpn. J. Appl. Phys.,* 41, L387 (2002).

[12] H. Matsuda, S. Okamura, T. Shiosaki, H. Adachi, and H. Sakakima, *J. Appl. Phys.,* 98, 063903 (2005).

[13] H. Matsuda and H. Sakakima, *J. Magn. Magn. Mater.*, 324, 1726 (2012).

[14] J. Mannhart, P. Chaudhari, *Physics Today* 54, 11 (2001).

[15] R. Mathieu, P. Svedlindh, R. A. Chakalov, and Z. G. Ivanov, *Phys. Rev. B* 62, 3333 (2000).

[16] J. M. D. Coey, A.E. Berkowitz, LI. Balcells, F. F. Putris, and F. T. Parker: *Appl. Phys. Lett.*, 73, 734 (1998).

[17] E. J. W. Verwey, *Nature* (London) 144, 327 (1939).

[18] T. Hibma, F. C. Voogt, and L. Niesen, *J. Appl. Phys.*, 85, 5291 (1999).

[19] H. Matsuda and H. Sakakima, *J. Phys.* D: *Appl. Phys.,* 44, 105001 (2011).

[20] H. Matsuda, abstract of 2nd International Conference of the Asian Union Magnetic Societies (ICAUMS 2012), Nara, Japan.

[21] M. Julliere, *Phys. Lett.*, 54A, 225 (1975).
[22] Gupta, J. Z. Sun, *J. Magn. Magn. Mater.*, 200, 24 (1999).
[23] Dieny, V. S. Speriosu, S. S. P. Parkin, B. A. Gurney, D. R. Wilhoit, and D. Mauri, *Phys. Rev. B* 43, 1297 (1991).
[24] S. Yuasa, T. Nagahama, A. Fukushima, Y. Suzuki, and K. Ando, *Nature Mater.* 3, 868 (2004).
[25] M. Bowen, M. Bibes, A. Barthelemy, J. –P. Contour, A. Anane, Y. Lemaitre, and A. Fert, *Appl. Phys. Lett.,* 82, 233 (2003).
[26] H. Sakakima, M. Satomi, Y. Sugita, Y. Kawawake, *J. Magn. Magn. Mater.*, 210, L20 (2000).
[27] H. Matsuda, Doctoral thesis, Nara Institute of Science and Technology, Japan, 2005.
[28] D. Lebeugle, A. Mougin, M. Viret, D. Colson, and L. Ranno, *Phys. Rev. Lett.,* 103, 257601 (2009).
[29] N. C. Koon, *Phys. Rev. Lett.*, 25, 4865 (1997).
[30] P. A. A. van der Heijden, C. H. W. Swüste, W. J. M. de Jonge, J. M. Gaines, J. T. W. M. van Eemeren, and K. M. Schep, *Phys. Rev. Lett.*, 82, 1020 (1999).
[31] Y. Ijiri, T. C. Schulthess, J. A. Borchers, P. J. van der Zaag, and R. W. Erwin, *Phys. Rev. Lett.*, 99, 147201 (2007).
[32] F. D. Saccone, P. Vavassori, and A. Berger, *IEEE Trans. Mag.,* 49, 4542 (2013).
[33] E. E. Fullerton, J. S. Jiang, and S. D. Baber, *J. Magn. Magn. Mater.*, 200 (1999) 392
[34] P. C. Dorsey, P. Lubitz, D. B. Chrisey, and J. S. Horwitz, *J. Appl. Phys.*, 79, 6338 (1996).
[35] L. Néel, *Comptes Rendus* 255, 1676 (1962).
[36] J. C. S. Kools, T. G. S. M. Rijks, A. E. M. Veirman, and R. Coehoorn, *IEEE. Trans. Magn.*, 31, 3918 (1995).
[37] J. G. Zhu and Y. Zheng, *IEEE Trans. Mag.,* 34, 1063 (1998).
[38] W. F. Brown Jr. *Micromagnetics*, (Harper and Row, New York 1963).
[39] The public domain OOMMF package is available at http://math.nist.gov/oommf/
[40] H. Takahashi, *J. Appl. Phys.*, 110, 123920 (2011).

In: Magnetite
Editor: Jacqueline Ward

ISBN: 978-1-63321-591-7

Chapter 4

MAGNETITE, PROPERTIES AND MODERN BIOMEDICAL APPLICATIONS

Marius Chirita[1,*], Mihaela Luminita Kiss[1,3] and Cecilia Savii[2]

[1]National Institute for Research and Development in Electrochemistry and Condensed Matter, Timisoara, Romania
[2]Institute of Chemistry Timisoara of Romanian Academy
[3]"Politehnica" University, Timisoara, Romania

ABSTRACT

This chapter outlines the fundamental properties of magnetite that makes it suitable for use in various biomedical applications. Magnetic properties of magnetite dispersed at nano and micro level are presented. It is well known that ferri or ferro magnetic materials with decreased particle size, at reduced nanometric range, become single domain and gain superparamagnetic behavior. In order to make use of these materials in biomedical applications, the particle agglomeration must be avoided after removing the polarizing field. This is the reason for presenting the theoretical concept of superparamagnetism. In addition, the efforts to extend the superparamagnetic behavior from nanometric level to

* Corresponding author: [1]National Institute for Research and Development in Electrochemistry and Condensed Matter, Timisoara, Plautius Andronescu Str. No. 1, RO-300224, Timisoara, Romania; chirifiz@gmail.com.

micrometric level by a different synthesis route will be presented. Extending this property from nanometric to micrometric scale is a real challenge and it may be very useful in the biomedical field, as microparticles are expected to be safer than nanoparticles due to their lower rate of diffusion in the tissue as well as to lower reactivity and weaker interaction with the immune system. A presentation of the main advantages and disadvantages of using ultrasmall (USPIO) or small (SPIO) nano-particles and micro-particles (MPIO) in different medical applications is underlined. In this chapter, our own results on the synthesis of single crystalline micrometric iron oxide with superparamagnetic behavior (generically named SCMPIO) will be presented along with proposed medical uses.

We will focus on applications of nano/micro magnetite particles in modern biomedicine (diagnostic and therapy technologies) such as: magnetic resonance contrast agents, drug delivery, gene delivery, hyperthermia for treatment of cancer, intracellular hyperthermia under high-frequency alternating magnetic field, cellular MRI, monitoring cell migration for cell therapy, multimodal cancer therapy, immunomagnetic separation of cells, detection, immobilization and modification of biologically active compounds, cell labeling, and magnetic separation of cells and bio molecule.

1. INTRODUCTION

Iron is a metal of the first transition series, placed in group VIII, period 4 of Mendeleev Periodic Table. It has 26 protons in the nucleus and atomic mass 56. It has four natural isotopes ^{54}Fe, ^{56}Fe, ^{57}Fe and ^{58}Fe and 10 synthetic isotopes. Iron exists in a wide range of oxidation states, −2 to +6, although +2 and +3 states are the most common, making combinations with oxygen O^{2-}. For example, ferrous oxide (FeO^- wüstite) is formed by accepting the electronegative oxygen O^{2-} of one ion Fe^{2+}. A combination of two ions, Fe^{3+} to three ions O^{2-} form Fe_2O_3 with (already known) four polymorphic forms alpha (hematite), beta, epsilon, and gamma (maghemite). The combination of four O^{2-} anions with divalent and trivalent Fe cations in 1/2 ratio is forming magnetite [1, 2, 3]. Among all magnetic iron oxides, magnetite shows the best magnetic properties with saturation magnetization greater than 90 emu / g; this property associated with its biocompatibility make magnetite useful in biomedical applications. The use of nano/micro magnetite particles in medicine is promising and attractive task due to the carrier potential under magnetic control and especially because of their potential to release the stored magnetic energy.

Taking into account the dimensions of the cell (10 μm - 100 μm), virus (20 nm - 450 nm), protein (5 nm - 50 nm), or gene (2 nm wide and 100 nm long), the choice of appropriate particle size is of utmost importance.

Three classes of iron oxides were used in medicine: *ultra small* (*USPIO*; 20 – 50 nm diameters), *small* (*SPIO*; 60 to approximately 250 nm), and micrometric (*MPIO*; 0.9 μm and upwards) [4, 5, 6].

While some biomedical applications require nanometer-sized particles, there are other ones require micrometer - size particles. For example, molecular MRI is performed by using targeted contrast agents that can alter either the T_1 or T_2/T_2^* relaxation of water protons near the target, thereby creating signal contrast. Gadolinium (Gd) chelates and superparamagnetic iron oxide nanoparticles (SPIO; 60-150 nm), (including the ultra-small - USPIO: 10-40 nm - and micron-sized versions - MPIO: 0.9-8 μm), are the two most popular classes of imaging agents that can generate contrast by shortening T_1 and T_2/T_2^*, respectively [7]. Any of those require specific magnetic properties: high saturation magnetization at a lowest possible intensity of the field and small remnant magnetization to avoid agglomeration after removal of the polarizing field.

Literature abounds in biomedical applications related to the use of particles of nanometric magnetite/maghemite because they exhibit superparamagnetic behavior only to a size smaller than a critical size (~15 nm). In order to exhibit superparamagnetic behavior, small iron oxide particles must be theoretically smaller than the magnetic domains.

Nevertheless, in biomedical applications, micrometer particles could bring a huge advantage compared to nanoparticles due to their lower tissue diffusion rate, lower reactivity, and lower interaction with the immune system [8]. The critical property required in case of micrometer-sized particles to be used in biomedical applications is the superparamagnetic behavior. Literature indicates [9, 10, 11, 12] that massive efforts were involved in the synthesis of single crystalline microparticles bearing superparamagnetic properties. To our best knowledge, most of the time the reported micrometric particles showing superparamagnetic behavior were actually generated by agglomeration of millions of nanoparticles, each of them having lower dimensions than magnetic domain have (10 nm). A disadvantage of magnetic microparticles built up as nanoparticle clusters is that their saturation magnetization is smaller compared to micrometric single crystalline particles of the same size.

There are relatively few reports [13, 14] in the literature which are focused on single crystalline magnetite micrometric particle synthesis. None of those actually even mention the superparamagnetic behavior of the obtained micrometric mono-crystals. Soft-magnetic crystalline magnetite octahedrons, with size distribution between 4 and 10 μm, have been hydrothermaly synthesized by Liu *et al.* [13]. The micro-crystals obtained by Liu *et al.* exhibit magnetic properties, as remanence ~6.6 emu/g and the saturation magnetization and the coercivity ~10 kOe and ~80 Oe respectively. Moreover, Liu *et al.* paper did not conclude on particle monodispersity.

Chen *et al.* [15] synthesised octahedral magnetite microcrystals ranging from 2 to 5 μm with high crystallinity and low coercive field. The magnetic hysteresis loop indicates the saturation magnetization as Ms = 102.57 emu/g, remnant magnetization as Mr = 1.44 emu/g, coercive field as Hc = 12.55 Oe, and reduced value of as Mr/Ms = 0.014 emu/g. Such material can be qualified as having very good magnetic characteristics. However, due to the high value of saturation (10 kOe), and the fact that particles are not monodisperse hinders their use for medical applications. Jiang *et al.* [16] developed a hydrothermal method to fabricate single crystalline pseudo-octahedral magnetite crystals ranging from 5 to 10 μm. Their coercivity was ~30 Oe, and at a saturation field of 10 kOe the saturation magnetization was 85 emu/g. The authors did not mention anything about remanence or monodipersity. Deng H. et al. obtained crystalline magnetite micrometer-sized particles with saturation at ~ 9 kOe, [17].

It is well known that materials with magnetic properties are significantly influenced by many factors such as: structure, (crystallinity, crystal defects), and morphology [13, 14]. The saturation magnetization of the magnetic particles increases with the particle size and the degree of crystallinity. On the other hand, the coercive force (Hc) increases with crystal defects number [14]. Consequently, in the case of micrometric crystallites, the effect of inside structure defects was expected to be more important when comparing surface properties.

In this context, our research group focused on developing a synthesis technique able to produce micrometric sized (between 0.5 and 10 μm) particles with single crystalline structure (not by clustering of superparamagnetic nanoparticles) having near superparamagnetic behavior (minimizing aggregation process in biologic media), genericly labeled as SCMPIO - Single Crystalline Micrometric Superparamagnetic Iron Oxide Particles.

In our research, the basic objective was to obtain microcrystals with as few as possible structural defects (the number of defects dramatically influences the magnetic properties). We inferred that reducing the crystal growth rate might alleviate the formation of defects in the crystal. Therefore, we chose the hydrothermal method and developed a new two-step synthesis procedure as follows: in the first step Fe-EDTA chelate coordination compound is obtained; in the second step, the chelate iron complex decay causes slow release of Fe^{3+} ions in the system. We reported [18] successful synthesis of single crystalline magnetite microparticles in the 1-45 micrometers range by hydrothermally decomposition of the Fe-Na_4EDTA complex, in the presence of urea. Our material presented high saturation magnetization (93 emu/g) at low intensity of the field (3.3 kOe), low remanence value (3.4 emu/g - 5 emu/g) and coercitivity (18.6 Oe – 26 Oe). These properties point to a superparamagnetic behavior, unusual in particles with micrometric size at room temperature [18, 19, 20]. Such particles could be fit for specific medical applications that require micrometer-sized mono-crystalline magnetite particles with near superparamagnetic behavior. The micrometer-sized particles synthesized by us present significantly higher saturation magnetization (near bulk theoretical one) when compared to clustered nanoparticulates. Moreover, due to their superparamagnetic behavior and their larger size, it makes them less susceptible to be collected in the blood stream (therefore increasing their bioavailability). In addition, due to their size, they exhibit weaker reactivity, and increased stability against biologic fluids (e.g. stomach acidic medium).

Therefore, improving the methods of synthesis of SCMPIO may entail tapping into new research directions in modern major themes or therapies. Procedures such as monitoring cell migration for cell therapy, multimodal cancer therapy, and targeting bimolecular therapy for a large class of different tissues illnesses, could be improved by using microparticles with superparamagnetic behavior. New suited facilities for bioinorganic chemistry and catalysis for environmental applications may emerge. They may also be successfully employed in applications such tissue drug delivery and/or contrast agent in the digestive tract. Main advantages and disadvantages of using USPIO, SPIO, MPIO and SCMPIO in certain medical applications (such as in vivo diagnostic, in vivo and/or in vitro therapy) will further be presented in this chapter.

2. Magnetite Physical Properties and Synthesis Route

Magnetite is black in color (Figure 1 a, b); it is one of iron oxides containing both Fe^{2+} and Fe^{3+} cations. The oxygen anions are overlapped in three layers ABCABC in a cubic close packed structure [21] with cell parameter a = 8.351 Å. The close packed structure means an arrangement infinitely dense of tangent spheres having the same radius, [22, 3] in which each sphere is in contact with 12 other spheres. The packing efficiency (volume of space occupied by the spheres/total volume) is 74.05 % which gives maximum density of the material. The iron cations are distributed in the tetrahedral and octahedral interstices generated by oxygen anions close-packed arrangement. This corresponds to an inverse spinel structure [23]. For a radius of the oxigen sphere 1, the distance from the centers of this sphere to the voids is for the tetrahedral arrangement and for the octahedral arrangement [22]. In this case, the minimum radius of the cations entering into tetrahedral voids is $\sqrt{2}-1$ and the minimum radius of the cations entering into octahedral voids is $\sqrt{\frac{3}{2}}-1$. If the radius of the cations is larger than this value (which is the case of the real spinel) the occupied sites expand on the expenses of free interstices. In the inverse spinel structure, 8 tetrahedral sites are occupied by Fe^{3+} ions and the octahedral sites are shared between equal amounts (8+8) of Fe^{3+} and Fe^{2+} ions. This distribution of cations creates two cationic lattices A (tetrahedral) and B[octahedral]. Under Curie temperature of 856° K- Nell temperature, the dipoles of the locations A and B are antiparallel which constitutes the source of ferrimagnetism of magnetite (Figure 1c). In other words, due to antiparallel arrangement of the magnetic dipole in octahedral and tetrahedral sites, the Fe^{2+} ions remain uncompensated, leading to ferrimagnetic behavior. The arrangement of magnetic dipoles can be described as $\uparrow Fe^{3+}$ [$^{\downarrow}Fe^{2+}\downarrow Fe^{3+}$] O_4. As temperature is increased towards the Curie point, the alignment (magnetization) within each domain decreases. Above the Curie temperature the magnetic domains disappear and magnetite becomes paramagnetic. Reducing the Curie temperature for nanoparticle versus bulk is linked to the degradation of the symmetry at the surface of the nanoparticles [2], and therefore the installation of a disorder in the orientation of magnetic dipoles [24]. Ultrafine magnetite particles with sizes less than 15 nm have superparamagnetic behavior, which means the magnetite is paramagnetic at room temperature [25], 26, 27, 28, 29].

At this dimensions, the thermal agitation at room temperature is strong enough to overcome the tendency of magnetic dipoles to be oriented. Bulk magnetite saturation magnetization value exceeds 90 emu/g. Magnetic properties of dispersed materials largely depend on the crystalline structure, synthesis method, particle size and shape.

Figure 1. a), b) SEM images of magnetite crystals; c) orientation of magnetic dipoles distributed in A and B sites.

When magnetite is thermally treated in vacuum or in an inert atmosphere, there is no oxidation and magnetic structure does not change. Exposed to thermal treatment in air, magnetite is successively transformed to maghemite and to hematite according to the scheme:

$$Fe_3O_4 \xrightarrow{< 300^\circ C} \gamma\text{-}Fe_2O_3 \xrightarrow{> 300^\circ C} \alpha\text{-}Fe_2O_3$$

Its black color and its band gap less than 2 eV enable magnetite to absorb the full visible spectrum (300 nm - 750 nm) radiation, which could be useful in photocatalytic and photoelectric applications. However, due to the fact that ionic radius of Fe^{2+} is longer than the Fe^{3+}, magnetite is susceptible to photo-dissolution process. Also, in time acid environment leads to the dissolution of magnetite.

In order to protect the nanoparticles against chemical degradation or agglomeration, it is often necessary to coat their surfaces with an organic or inorganic shell [30, 31, 32, 33]. The shell is developed and evaluated silica-based micron-sized iron oxide particles (sMPIOs) with a functionalizable particle surface. Coating can also be performed in order to add new functionalities to the magnetic core, such as biological stealth, optical properties, catalytic or adsorbing capacity. The core-shell and core-shell–shell structures and nanocomposite structures are particularly interesting architectures in photocatalysts. Numerous papers describe magnetic nanoparticles coating with an organic polymer [34, 35, 36, 13], or a silica shell [37, 38, 39, 40] and relatively few investigations have been performed on magnetic nanoparticles coated with titanium dioxide (TiO_2) [38, 41, 42, 43, 44, 45, 46, 47, 48, 49] that can be used in light-assisted environmental cleanup (photocatalysis) [50], or in solar energy conversion [51]. Therefore, the addition of a TiO_2 shell to a magnetic core at the nanometer scale may lead to bifunctional nanoparticles which could be applied for instance as magnetic photocatalyst. Papers reporting TiO_2 coated with magnetite (Fe_3O_4), maghemite (Fe_2O_3), or ferrite (MFe_2O_4) - where M is a divalent cation [38], [44, 47, 52] - describe rather large or aggregated particles. As magnetic photocatalysts, the efficacies of these materials are generally low. This was explained by an unfavorable electronic interaction between the TiO_2 shell and the magnetic core, leading to an increase in electron-hole recombination [38]. To avoid this problem, several authors [41, 42, 43, 44, 49] synthesized $Fe_3O_4/SiO_2/TiO_2$ and $BaFe_2O_4/SiO_2/TiO_2$ particles, in order to isolate TiO_2 from the magnetic core by a passive silica layer. Although activities of the magnetic photocatalysts were increased, they actually remain inferior to those of the commercial TiO_2 photocatalyst. This can be explained by reduced accessibility to catalytic sites of obtained large particles or aggregates having diameters of more than 500 nm.

As shown, while the magnetic properties of magnetite can be used in applications related to solar energy conversion, the main applications of magnetite are in the storage information domain, and in the biomedical field. In the field of information storage, materials were needed to have a good memory, specifically to have a high level of remnant magnetism. In contrast, medical applications require particles with small or zero remnant magnetism, in order to avoid aggregation due to magnetic attraction. In other terms, this means they need superparamagnetic behavior. Moreover, given that organic tissue types diverge (healthy tissue, ill tissue, benign tumors, and malign tumors) their chemical loads have to differ substantially from one to another.

There are two fundamental approaches to fabricating nanomaterials. The "bottom-up" approach allows of constructing a nanomaterial from basic building blocks. This approach illustrates the possibility of creating exact materials that are designed to have exactly the desired properties [53]. In order to create a nanostructured material, the "top-down" method proceeds by restructuring of a bulk material [54]. Top-down method is usually not very well suited to preparing informally shaped particles or very small sizes. Bottom-up procedures are much better suited to generate uniform particles, often of distinct size, shape, and structure [55].

Preparation of nanomaterials can be as well classified into physical and chemical methods. It is notable that some methods can be considered as either chemical or physical routes depending on the media, precursors, and operating conditions. Various methods to produce nanoparticles can be classified based on the process media, including vapor, liquid, and solid state processing routes, and combined methods, such as vapor—solid—liquid approach [56]. The production of nanoparticles involves techniques to hinder the natural course of thermodynamics through manipulation of kinetics, which is the case of sol-gel method of synthesis. In other cases it is possible to hinder the natural growth of phases through the use of dilution or via protection of surfaces using surface-active agents or by coating and encapsulation of nanoparticles in a glassy media, hybrids or polymers.

The synthesis of magnetic nanoparticles has been an area of study for a long time because of the interesting practical applications of such particles. Various techniques, but most often chemical synthesis methods for preparing magnetite nanoparticles, have already been reported, [57]such as coprecipitation [58] (Massart's method; coprecipitation of Fe^{+2} and Fe^{+3} salts in saturated base solutions [59], [60] microemulsions [61], solvothermal processing [62] and high-temperature organic phase decomposition [63, 64, 65] Andrade *et al.* have prepared size-controlled superparamagnetic iron oxides nanoparticles through a reduction–precipitation method combined with a thermal treatment.

Magnetic nanoparticles were prepared and frequently tested in various formulations for biomedical applications. Synthesis methods and studied parameters have been closely related to the application in order to allow for the development of optimized characteristics and functionalities suitable for target goal.

Multifunctional nanocomposites combining with magnetism of magnetic particles and physicochemical properties of functionalized coatings have become ideal candidates for applications in the field of biomedicine. Countless publications deal with the subject, which demonstrates its importance. Moreover, there are many issues to be solved in the field.

Some of last year's publications we found describe the synthesis methods of magnetic iron oxide and composite particles closely correlated with the applications they are intended to be used for. For instance, synthesis and formulation, together with physical-chemical characterization and in biologic media behavior assessment of magnetic materials could have been found in cell labeling/cell separation [66, 67] drug delivery conjugated to the coated dextran or other polymeric coatings such as starch, polyethylene glycol, block copolymers, [68, 69, 70]. MNPs coated with oleic acid (OA) and then stabilized with pluronic F-127 to form a water-dispersible system [71], gene delivery [72], contrast agents for magnetic resonance imaging (MRI) [73], MRI contrast [74] to induce local hyperthermia in response to an external alternating magnetic field to selectively destroy cancer cells [75] as a magnetically targeted carrier system in drug delivery applications [68], [76] are good examples of specific applications of nanoparticles. Zang and coworkers obtained multifunctional magnetic nanocomposites having spherical shapes (with the mean diameter of 180 nm) exhibiting good dispersion and hydrophilicity by assembling multifunctional nanocomposites with iron oxide and amino-terminated PAMAM dendrimers [77, 78]. MNPs can be used in conjunction with drug delivery applications for real-time monitoring of drug distribution to the target tissue, as well as for following the effect of therapeutics on the progression of disease [79]. Jain and coworkers demonstrated that: a single drug or combination of drugs can be incorporated in our MNPs with high efficiency; the combination of drugs in MNPs demonstrates synergistic antiproliferative activity in cancer cells. In addition, the incorporated drugs do not adversely influence the magnetic and imaging properties of the formulation [73].

2.1. Magnetic Particulate Systems Suitable for Biomedical Applications

2.1.1. Nanoparticles

Iron is a very important metal in living systems, especially for oxygen transport objectives and electron transport, linking the oxidation of substrates to the reduction of O_2. Generally, iron can link any kind of biomolecules. Therefore, the iron can adhere to membranes, proteins, nucleic acids. While "free" iron (which binds non-specifically to many cellular components) can catalyze the production of toxic free radicals, in cells, iron is generally stored in the center of metalloproteins. In nature, many redox reactions are dependent on iron-containing enzymes whereby electron transport is facilitated by changes in the metal oxidation state [25]. Nitrogen fixation and photosynthesis are examples of processes in which iron-containing enzymes play vital roles.

Many neurodegenerative diseases indicate the disruption of normal iron homeostasis in the brain. Recent experimental work indicates that nanoscale magnetic bio-minerals (primarily magnetite and maghemite) may be associated with senile plaques and fan filaments found in brain tissue affected by these diseases. Understanding the role of iron in neurodegenerative disease could help understand the origin, diagnosis and treatment of these diseases [80]. A study [81] of the magnetic properties of superior temporal brain tissue from Alzheimer's disease (AD) and age-matched control subjects demonstrates an exponential correlation among the concentrations of the Fe^{2+} ion-containing iron oxide, magnetite- Fe_3O_4, and the fraction of those particles that are smaller than 20 nm in diameter.

These data provide circumstantial evidence in favor of biogenic synthesis of 8 nm diameter iron particles and their storage in protein ferritin. For the first time the authors also show that the total concentration of biogenic magnetite is generally higher in the AD brain (in some cases as much as 15 times greater than controls) and there are gender-based differences, with AD female subjects having significantly higher concentrations than all other groups. These results provide insights which may guide current efforts to develop iron-based MRI diagnosis of AD.

From this point of view, especially for their magnetic properties and for their biocompatibility, the ferromagnetic and superparamagnetic nanoparticles Fe_3O_4 are involved in many interesting applications in biology and medicine. Among the key properties of the nanoparticles involved in bio-applications was particle size, and particle size distribution. The tiny particles (below one micrometer) have to penetrate the tissue and organs at cellular and subcellular level. Also, the monodispersity allows foreseeing the particle collection behavior equally obeying the physics and chemistry laws inside the living bodies. In order to make inorganic particles biocompatible, they need to be coated with biopolymers that make bigger functionalized particles. The optimal magnetic core of functionalized nanoparticle is 20-50% from the total weight [25]. Typically, magnetic nanoparticle probes for biomedical applications are comprised of nanoscale superparamagnetic iron oxide (SPIO) cores of magnetite and/or maghemite encased in polysaccharide, synthetic polymer, or monomer coatings. Due to the Si-OH group on the silica surface, silica coated nanoparticles have also a great potential for biomedical applications.

Paramagnetic lanthanides with unpaired electrons are usually used as contrast agents in MRI applications. Most of the metals with unpaired electrons are not suitable for clinical MRI because of the unmatched metal spin relaxation time and proton Larmor frequency. The rare earth elements with suitable characteristics include gadolinium (Gd^{3+}), and dysprosium (Dy^{3+}), beside transitional metals, manganese (Mn^{2+}) and iron (Fe^{3+}) [82]. The number of the unpaired electrons directly correlates with the strength of the magnetic moment of the element. Gd^{3+} has seven unpaired electrons and long electronic relaxation time, hence a strong magnetic moment. This property of Gd^{3+} makes it an excellent paramagnetic contrast agent. Dy^{3+}, Mn^{2+} and Fe^{3+} have five unpaired electrons. Hydrogen, the natural source of the MR signal, has only one unpaired electron. The superparamagnetic behaviour of the subdomain magnetic cores (3–10 nm) is similar to that of paramagnetic substances; they lose their magnetization when the magnetic field is removed, but differs by the value of the magnetic moment, which is markedly higher for the superparamagnetic ones. Therefore superparamagnetic materials relaxivity values are much higher than those of the widely-used paramagnetic Gd-chelates. Often iron oxide based superparamagnetic materials are used for their capacity to produce T_2 relaxation effects, which result in signal reduction on T_2-weighted images. They are generally synthesized in a one-step process by alkaline coprecipitation of iron (II) and iron (III) precursors in aqueous solutions of hydrophilic macromolecules, such as dextran, followed by chromatographic separation for narrowing the polydispersity domain. Dextran macromolecules serve to limit the magnetic core growth during the synthesis and to stabilize via sterical repulsions the nanoparticle dispersion in water (and later in physiological medium). These colloidal contrast agents would be more realistically described as several magnetic cores, more or less aggregated, embedded in the dextran corona, which are sometimes cross-linked in a second step for enhancing the mechanical entrapment of the inorganic cores [83].

SPIO typically consists of two components: an iron oxide core and a hydrophilic coating. Coatings can improve oxidation resistance, colloidal stability, the ability to functionalize, phagocyte resistance, mechanical stability, and biocompatibility [84]. It is well established that gold coating offers corrosion resistance, even if biotoxicity issues may need to be solved. Polysaccharide coatings such as dextran, starch, and chitosan are biocompatible and offer a range of functionalization options [85, 86, 87]; however, they can be structurally weak and can be dissolved by highly acidic environments. Silicon-based coatings are used to protect particles from lysomal enzymatic digestion and to improve mechanical properties and chemical stability [88, 89, 90]. Silica-coating can improve chemical stability, but a porous coating may allow the contents inside to be dissolved or oxidized [91, 92]. Many polymers are biocompatible and may be used as coatings for metallic or ceramic particles or can serve as hosts by either capturing nanoparticles inside a larger polymer particle or by attaching nanoparticles to their surfaces. Polyethylene Glycol (PEG) and related polymers covalently bond to surfaces or are adsorbed on magnetic nanoparticles and can prolong circulation time in the bloodstream [85, 86].

A very good classification of the SPIO sizes in literature is presented by Andrew Tsourkas and coworkers in [93]. SPIO sizes range greatly from 2 to 3 nm for citrate inhibited growth SPIO, tens of nanometers for polymer coated polycrystalline iron oxide nanoparticles through to micrometers can be used for orally ingestible contrast agents. Larger diameters are available and are useful in such enterprises as cell tracking and separation, cell rheology and membrane deformation, and as contrast agents for the gastrointestinal tract. However, they have limited functionality in molecular imaging applications due to their limited accessibility to neo- and microvasculature. Categories of SPIO, based on their overall diameter (including iron oxide core and hydrated coating), are noted in the literature [94] as oral-SPIO between 300 nm and 3.5 μm; standard SPIO (SSPIO) at approximately 60–150 nm; ultra-small SPIO (USPIO) of approximately 10–40 nm [95]; and nanocrystalline iron oxide nanoparticles (MION—a subset of USPIO) of approximately 10–30 nm [96]. MION nanoparticles are so named to underline the single crystal nature of their core, in contrast to SPIO greater than 50 nm that are comprised of multiple iron oxide crystals.

Because of their large overall hydrodynamic volume (over 50nm in diameter), SPIO agents are efficiently accumulated in the organs of the Mononuclear Phagocyte System (MPS): ca. 80 – 90% of the injected dose in liver and 5–8% in the spleen with plasma half-life less than 10 min [97]. Therefore, SPIO decrease liver and spleen signals and allow for diagnosing malignant tumors or metastases in these organs. Thanks to their smaller size and to the hydrophilicity of their dextran corona, USPIO act as stealth particles. Their plasma half-life is more than 2 h [98] and therefore they remain in the blood long enough to act as blood-pool agents (MR angiography) [99, 100]. This size-dependent distribution in tissues (passive targeting) is a current limitation for the diagnosis of pathologies in any other organ. For instance, imaging specific tissues would need contrast agents as stealthy as USPIO with an extra requirement: their surface should be labeling with ligands that specifically bind to surface epitopes or receptors on the target sites (active targeting). Only very few attempts for conjugating biomolecules to the dextran coating were reported using electrostatic or chelating interactions [101] or reduction of Schiff's bases formed from dextran activated by oxidation [102, 103]. The interactions between magnetic cores and dextran macromolecules, e.g. Van der Waals and hydrogen interactions [104], are too weak and generally prevent any efficient derivatization of dextran corona without macromolecule depletion [105]. Recently, for a similar purpose, attempts were made for preparing USPIO from dextran whose terminal sugar had been previously reduced [106]. In [107] an investigation for a new generation of USPIO is presented, based on maghemite cores covalently bonded to their dextran corona through the use of silane coupling agents.

Because of their potential ability to be tailor-derivatized, they are called VUSPIO for Versatile USPIO.

2.1.2. Microparticles Composed of Clusters of Nanoparticles

USPIO and SPIO are frequently recommended and used as contrast agents in MRI in order to increase both magnetization and transverse (T_2) relaxivity [108]. They present limitations due to their fine particle size. In an attempt to overcome this aspect two ways were envisaged: first, by increasing every single particle dimension and second, by clustering individual SPION into clusters [109]. Bean and Livingston and others showed that the large particles (at a size approximately larger than 15 nm) are not superparamagnetic [110] and easily aggregate in solution. However, using block copolymers template Berret et al. [111] fabricated maghemite nanocluster and tuned the size of aggregates in the range of 70–150 nm with aggregation numbers (number of nanoparticles per aggregate) from tens to hundreds. In this way the transverse relaxivity, was noticeably increased with the size of the magnetic clusters. Notable, in case of clusterization the specific saturation magnetization (Ms) values of nanoparticle cluster, corresponding to the complete alignment of all individual moments in a sample, are smaller than their corresponding to bulk phases due to disordered crystal structure which increases with particle size reduction [112].

Synthesis by co-precipitation was reported of a magnetic system consisting of agglomerations of magnetite nanoparticles with superpara-magnetic core (11.8 μm) and covered by amoxicillin.

This has been proposed for use in the treatment of gram negative bacteria Helicobacter pylori spiral shape [8]. Numerous studies [113, 114, 115] are focused on applications of MRI in molecular and cellular imaging. Hence, obtaining contrast agents with small relaxation times which can improve MRI signal are of high interest. It has been shown that nanoparticles of USPIO type increase intensity of MRI signal in receptor detection [116, 117] as well as monitoring cell migration [118, 119].

In order to increase the efficiency of the MRI signal, the above listed research results show that it is necessary to have several million for USPIO nanoparticles to mark a single cell. This is a challenging task [120]. Hinds et al. [121] showed that alternately, micrometric particles MPIO can be effectively endocitated by various cells and can thus contribute to the improvement of MRI signal. These particles can be coated with a polymer and may be labeled with a fluorescent agent; they can be used both for the improvement of MRI imaging and for tracking cells by fluorescence microscopy.

Shapiro et al. [120], showed that MPIO carry iron an order of magnitude greater than USPIO and can cause a local magnetic field inhomogeneity extending for a distance approximately 50 times the physical diameter of the particle. Shapiro's experiments on MRI resolution effects indicated that T2* effects can be readily detected from single MPIOs at 50-μm resolution [120]. In 2007, researchers group led by Robin Choudhury at Oxford University, England, have developed a new magnetic resonance imaging method that could help with the diagnosis and management of multiple sclerosis. The technique uses antibodies tagged with microparticles of iron oxide (MPIO) to detect expression of VCAM (endothelial vascular cell adhesion molecule) in blood vessels of the mouse brain. The authors suggest the new method could be adapted to monitor other endovascular molecules implicated in different pathologies; it also has the huge advantage that it can detect inflammation long before other multiple sclerosis (MS) pathologies are apparent [122]. Later on, in 2011 Choudhury and co-workers continued their studies concerning MPIO potency as contrast agent in medical imaging techniques. They reported that for molecular magnetic resonance imaging (mMRI), microparticles of iron oxide (MPIO) create potent hypointense contrast effects especially due to their physical size.

Several MPIO characteristics that make it promising contrast agent were underlined:

- MPIO has higher content of iron oxide, evaluated at approximately one order of magnitude higher than that contained in ultrasmall particles of iron oxide (USPIO) commonly used for MRI contrast;
- Consequently, due to great iron content, MPIOs create potent negative contrast effects on T_2*-weighted images that extend to a considerable distance;
- The contrast MPIO potency enabled in-vivo single MPIO-labelled cells detection and tracking;
- Due to their size, MPIOs are less susceptible than USPIOs to nonspecific uptake by endothelial cells so that the microparticles retain specificity for endovascular molecular targets;

MPIOs are commercially available with a range of reactive groups, providing the opportunity for covalent conjugation of biomolecules, but Mankia et al., showed that MPIOs used so far in experimental studies have a nonbiodegradable polyurethane coat, and are therefore not suitable for human use. For the purposes of clinical translation, there is ongoing work to develop biodegradable MPIOs by modification of the surface coat [123]. In certain cases, when other MRI techniques were inefficient, the solution could be found by using conjugate MPIOs. In mouse brain inflammatory pathology MPIOs of about 1μm was the used contrast agent which enabled the disease to be in vivo detected. MRI of the brain was performed at 7 Tesla using a T_2*-weighted imaging [123].

Due to their size and incompressible nature, MPIOs are less susceptible to nonspecific vascular egress or uptake by endothelial cells. A novel, functional MPIOs probe targeting vascular cell adhesion molecule-1 (VCAM-1) for detection of acute brain inflammation in vivo at a time when pathology is undetectable by conventional MRI was developed [123]. The application field of MRI micrometric magnetic contrast agents is in expansion. Several applications are mentioned by Mankia et al. Larger microparticles of iron oxide (MPIO) have been used to good effect for MRI in cellular imaging [124], characterization of vascular inflammatory disease, [125, 126] molecular imaging of thrombosis, [9], molecular imaging of tissue ischemia [10]; MPIOs may be a useful contrast agents for detection of endovascular molecular targets by MRI [11].

Speaking about MPIOs biocompatibility and toxicity, [123] mentioned that the doses of MPIOs used in animal studies have not caused tissue infarction, thrombosis, or vessel plugging in vivo, and animals have not shown any ill effects in the short term. MPIOs are cleared far more rapidly from the circulation than USPIOs by liver and spleen.

Concerning clinical translation, MPIOs are not yet so far in human use [123]. For the purposes of clinical translation, there is ongoing work to develop biodegradable MPIOs by modification of the surface coat which is still a barrier for clinical application [123], [127].

Oxide magnetic particles suspension in distilled water with average size particles of 10.82 μm were used as MRI contrast agents by Mathieu and Martel [128]. The magnetite weight fraction inside the particles was above 90%, corresponding to a minimum particle (significant) magnetic moment of 82.8 emu/g taking (for bulk magnetite, Msat = 92 emu/g). Mathieu and Martel is also mentioning MRI imaging, by using millimetre-size particles in larger, and in smaller vessels [128]. Little was found about successful use of contrast agents in MRI investigations of gastrointestinal tract, especially concerning orally administrated ones.

2.1.3. Single Crystalline Micrometric Iron Oxide with Superparamagnetic Behavior

A viable alternative for enhancing the MRI contrast is to develop micrometric monocrystalline particle as magnetic core contrast agent.

As previously mentioned, the iron oxide particles with dimensions below 15 nanometers often exhibit superparamagnetic relaxation (i.e., fluctuations of the magnetization vector in the proximity of the easy axe of magnetization). In a series of applications, a major advantage of the superparamagnetic behavior consists in the fact that after the polarizing field removal particles agglomeration is avoided. Extending this advantage from nanometric to micrometric scale it is a real challenge and could be very useful. In the biomedical field, due to the lower rate of diffusion in the tissue as well as due to lower reactivity and weaker interaction with the immune system, the use of microparticles is expected to be safer compared to the use of nanoparticles. Great efforts in order to synthesize micrometric particles have been made lately. Nevertheless, there are currently few reports on the synthesis of magnetite at micrometric scale. None of these reports was specifically citing the synthesis of superparamagnetic microparticles with single crystalline structure. All the biological research has been performed on micrometer magnetite comprising nanoparticle agglomerations. Due to its compact nature, single crystalline micrometric magnetite produces much larger magnetic effects compared to nanoparticle agglomerations. The high value of saturation magnetization determined by single crystalline compact structure if it is associated with very low value of remanence could be very useful for some biomedical applications. It was estimated that crystalline micrometric particles with superparamagnetic properties could considerably enhance MRI contrast effects. Also, the results regarding synthesis of SCMPIOs could open new perspectives in medicine such as in: intracellular hyperthermia under high-frequency alternating magnetic field for treatment of cancer, controlled drug delivery system (site specific drug delivery), cellular MRI, monitoring cell migration for cell therapy, multimodal cancer therapy, immunomagnetic separation of cells, detection, immobilization and modification of biologically active compounds, cell labelling, magnetic separation of cells and bio molecules, multi-modal cancer therapy. In this context, we attempted developing a new class of magnetic material, specifically micrometric single crystalline magnetite particles with superparamagnetic behaviour as MRI T_2 contrast agents, named as SCMPIO. By using a new synthesis procedure, hydrothermal decomposition of the Fe-EDTA complex, we developed an inspired route to synthesize (1-50 μm) micrometric single crystalline magnetite exhibiting superparamagnetic behaviour at room temperature [18, [19, 20, 129, 130]. We obtained microcrystals with 10-15 μm in size [18] at 250º C after 2 h of high temperature treatment time. Otherwise we obtained microcrystals having 45μm in size (measured along the <001> axis) [19] at 230° C, after 38 h of high temperature treatment time with morphology showing a combination of octahedral and dodecahedral faces. In [129] we presented an interesting transition from single phase of hematite (after the first 4 h) to single phase of magnetite microoctahedrons (after 40 h) with superparamagnetic behaviour at room temperature via single phase of iron carbonate microcrystals. Scanning Electron Microscopy (SEM) and X-ray Diffraction (XRD) were performed (Figure 2 a and b) to examine the final product. As the magnetic measurements have shown (Figure 2 c) the magnetite synthesized by this procedure has low remnant magnetization (3.4 emu/g), less than 3.5% from saturation and coercivity (18.6 Oe).

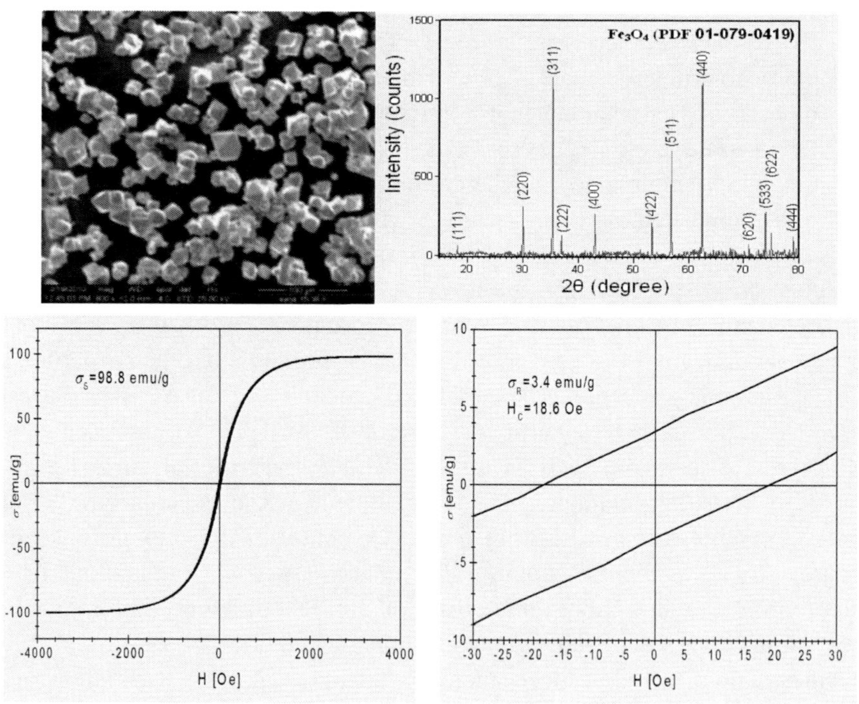

Figure 2 a: SEM image b: XRD spectrum c: cycle of magnetization d: detailed.

The crystals show at room temperature a magnetic behavior very close to superparamagnetism, which is unusual for single crystalline particles of size in the micrometric range.

The promising suitability of micrometric single crystalline superparamagnetic magnetite as MRI contrast agent was mainly based on previous results of de Hinds and Shapiro. Hinds et al. [121] showed that MPIOs micrometer particles may be efficiently endocited by various cells and may thus contribute to improved MRI signal. These particles coated with a polymer and marked with a fluorescent agent can be used both to improve MRI imaging and cell tracking by fluorescence microscopy. Shapiro in [120] demonstrated for the first time that a single MPIO particle can be used for marking one cell and then detected by MRI. Shapiro placed such particles in an embryo and demonstrated that these particles allow tracking the cells through MRI imaging and this was proved by histological examination. On the other hand, stem cells have a high proliferation rate; marking them with nanoparticles SPIO lessens the MRI signal with the number of cell divisions (because nanoparticles are distributed to daughter cells, and the signal intensity decreases). Hence, marking one micrometer particle/stem cell tracking by MRI would allow their migration to the area of interest. The ability to detect a single particle in one cell by MRI is a great opportunity for cellular imaging.

Even millimetre-size magnetic particles were mentioned in literature as accessible as MRI contrast agents. Hence, there are no limits in the nm to mm size range for investigating the field. As mentioned previously there are already many applications where micrometric magnetic particles have demonstrated their capabilities as MRI agents. All the studies clearly show that in order to improve MRI signal quality for marking a single cell micrometric particle made by aggregation of millions of USPIO nanoparticles was exclusively used [120]. Our preliminary tests have shown that SCMPIOs generates an appropriate magnetic MRI response at 10^{-4} g/cm^3 concentration [131].

3. Biomedical Applications

There is a large amount of literature on magnetic materials and on iron oxides specifically. Designing iron oxides with set magnetic properties, especially superparamagnetic behavior, is of great interest for eclectic biomedical applications. Special attention is paid to their specific design and synthesis, in order to suite well the targeted application. A brief description of selected biomedical applications will follow. This includes iron oxides as MRI contrast agents, drug delivery systems, gene delivery, multimodal cancer therapy, hyperthermia for treatment of cancer, intracellular hyperthermia under high-frequency alternating magnetic field, cell labeling, and cell separation.

3.1. MRI Contrast Agents

3.1.1. The MRI Basics

The possibility to create a two-dimensional picture by introducing gradients in the magnetic field was discovered by Paul C. Lauterbur (born 1929), Urbana, Illinois, USA. In 1973, he described how addition of gradient magnets to the main magnet made it possible to visualize a cross section of tubes with ordinary water surrounded by heavy water. Further, Peter Mansfield (born 1933), Nottingham, England, developed the utilization of gradients in the magnetic field. He described how the signals could be mathematically analysed to develop a useful imaging technique. Mansfield also showed how very rapid imaging could be achieved by very fast gradient variations (so called echo-planar scanning). For their descoveries, Paul C. Lauterbur and Sir Peter Mansfield was rewarded with Nobel Prize in Physiology or Medicine in 2003. Much earlier, based on the resonance principle first described by Irish physicist Joseph Larmor, Isidor Isaac Rabi developed a technique for measuring the magnetic characteristics of atomic nuclei. He's technique enabled more precise measurements of nuclear magnetic moments. He won the Nobel Prize in Physics in 1944. Later, Edward Purcell and Felix Bloch laid the foundations for magnetic resonance imaging (MRI) by improving the Rabi's method. In 1952, for their work they were awarded with Nobel Prize in Physics. Another Nobel Prize was awarded to Swiss scientist Richard Robert Ernst in 1991 for contributions to the development of the method of high-resolution nuclear magnetic resonance (NMR) spectroscopy. Ernst and a colleague found that NMR spectroscopy could be more effective if the slow sweeping radio waves traditionally used to bombard a sample were replaced by short, intense pulses. They improved the sensitivity by as much as 100-fold by using a computer to perform a complex series of mathematical operations (Fourier transformations) in the received signal; the method allows the study of small amounts of material and additional types of nuclei.

The basic principle of MRI is based on nuclear magnetic resonance (NMR) together with the relaxation of proton spins in a magnetic field.

Relaxivity is a common parameter, considered important, among others, exploited in the generation of MRI contrast. Runge explain [4] in accessible way the significance of relaxivity parameters. There are two relaxivity parameters that are unique to each tissue: T_1 and T_2. Longitudinal or spin-lattice relaxation time, known as T_1, refers to the amount of time it takes for the tissue magnetization to return to its equilibrium state in the longitudinal direction of the main magnetic field after excitation with a radiofrequency (RF) pulse. The excess energy that is absorbed by the magnetic spins from the RF pulse is transferred back to the environment during the relaxation process. The second relaxivity property of tissue is transverse or spin-spin relaxation, referred to as T_2 relaxation. In this relaxation process, the excess energy deposited in the tissue by the RF pulse is subsequently transferred between magnetic spins. This transferred energy results in loss of spin phase coherency in the transverse plane and spin dephasing.

Once established scientific and technical operating conditions of MRI equipment for medicine, an area of maximum interest was developed for obtaining MRI images as best as possible (best contrast, best resolution).

In this respect there is need for more efficient synthetic contrasts agents based on nanometric superparamagnetic particles are needed to be infused in biological tissues. The MR contrast agents are made of magnetic nanoparticles and represent a heterogeneous group of pharmacological structures. Basic principles of MRI contrast agents were described by Bin Na and other researchers; they showed that T_1 contrast agents are usually paramagnetic complexes, while T_2 contrast agents are based on iron oxide nanoparticles, which are the most representative nanoparticulate agents [132]. They give an intuitive expression of what really means the contrast term: 'Contrast' refers to the signal differences between adjacent regions, which could be 'tissue and tissue', 'tissue and vessel', and 'tissue and bone'. The delivery of nanoparticles in the human body usually requires suspending the nanoparticles in a water-based fluid, and must be stabilized against aggregation due to Van der Waals or magnetic interactions.

There are different classification methods, but the most commonly used criterion was based on contrast agent biodistribution in tissues: extracellular, intracellular, tissue specific, and blood pool/intravascular contrast agent [82]. A good intracellular contrast agent is manganese (Mn^{2+}), an essential trace metal of human diet and a powerful paramagnetic metal with five unpaired electrons. Its free form is toxic if directly administered to circulation. To minimize the potential toxicity and increase tolerance, Mn^{2+} must be chelated. For instance, in commercially available forms, contrast agent delivered intravenously to enhance contrast, named mangafodipir trisodium (Mn-DPDP), or mixed with calcium ions.

A good indicator of some metabolic disorders and disease states is the tissue specific contrast agent, based on metal-porphyrins. Their metal-chelated forms show selective retention in tumor tissue. Studies in myocardium have drawn attention to possible future use. Specifically, metalloporphyrins have demarcated irreversibly infracted myocardium from viable myocardium.

To our best knowledge, at present day, no human intravascular/blood pool agents is currently FDA approval. The first blood-pool agent (ABLAVAR®) was only recently FDA approved for MRA for evaluation of aortoiliac occlusive disease in patients with suspected peripheral vascular disease. The applicability of blood-pool agents in various other indications is currently being assessed in clinical trials [133] investigation [82].

A very good candidate as contrast agent was superparamagnetic iron oxide nanoparticles (SPIO). The agent is composed of a water-insoluble crystalline magnetic core (magnetite or maghemite) and dextran or starch derivate coat. All iron oxide particles for intravenous injection first distribute into the intravascular space as blood pool agents. They are subsequently internalized and degraded by the macrophages in liver, lymph nodes, and bones and enter the total iron body storage after degradation [134]. Particles introduced into the bloodstream are covered rapidly by components of the circulation, such as plasma proteins, in a process called opsonization. Antibody opsonization is the process by which a pathogen is marked for ingestion and destruction by a phagocyte. Opsonization involves the binding of an opsonin, i.e., antibody, to a receptor on the pathogen's cell membrane. Opsonization makes the particles recognizable to the body's major defense system, the reticuloendothelial system (RES). The RES comprises a diffuse system of phagocytic cells (which engulf inert material) that are primarily associated with the connective tissues in the liver, spleen, and lymph nodes [84]. Macrophage (Kupffer) cells in the liver and macrophages of the spleen and circulation are important in removing particles identified by opsonization. A significant fraction of nanoparticles can be cleared from the circulation system in as little as 15 minutes [135, 136], the clearance rate depending on size, charge, surface hydrophobicity and the number and nature of functional groups on the surface. Smaller particles more easily escape from the RES, but the smaller size usually has a smaller moment. Anionic particles with negative surface charge have a high affinity for the cell membrane and are typically taken up by the endocytic process. Cationic magnetite particles induce significantly lower cell-survival; their toxicity depends highly on the magnitude of the surface charge [84]. More highly cationic particles tend to be more toxic. Hydrophilic surfaces such as dextran, polyethylene glycol, polyethylene oxide, poloxamers, polysorbates and polyoxamines provide a dynamic 'cloud' of hydrophilic and neutral chains at the particle surface that repel plasma proteins and prevent rapid removal of particles from circulation. Finally, the iron oxides are metabolized into elemental iron and oxygen by hydrolytic enzymes in the body. The iron joins the normal body stores and the body compensates by taking up less iron from the stomach [84].

3.1.2. MRI Contrast Enhancement. Contrast Agents

Usually, considering their size, three classes of iron oxides were used in MRI: ultra-small particles of iron oxide (USPIO; 20–50 nm diameters), small particles of iron oxide (SPIO; 60 to approximately 250 nm), and microparticles of iron oxide (MPIO; 0.9 μm and upwards) [4, 5], [108].

Gadolinium chelates and superparamagnetic SPIO, including USPIO and MPIO, are the two most popular classes of imaging agents that can generate contrast by shortening T_1 and T_2/T_2^*, respectively [7].

Basic criteria were identified that need to be met in the design of an MRI contrast agent, such as:

- to alter the parameters responsible for image contrast at low concentrations to minimize dose and potential toxicity;
- to possess some tissue specificity in vivo so that it is delivered to a tissue or organ in a higher concentration than to other areas in the body and to remain localized for a reasonable period of time so that imaging can be performed;
- to be substantially cleared from the targeted tissue or organ in a reasonable period
- of time, usually several hours after imaging;
- to minimize potential effects from chronic toxicity. The contrast agent must also be excreted from the body, usually by renal or hepatobiliary routes;
- to have low toxicity and to be stable in vivo while being administered in doses that can affect the MRI relaxation parameters sufficiently to result in visible contrast enhancement. The dose levels of the contrast agent required to meet these criteria must be evaluated by several tests depending on intended use;

- to possess a suitable shelf life for storage. It must remain stable in vitro for a reasonable period of time while being stored. A shelf life of years is desirable.

At small sizes (on the order of tens of nanometers), ferri- or ferromagnetic materials, such as magnetic nanoparticles, MNPs, become a single magnetic domain maintaining one large magnetic moment. At sufficiently high temperatures (i.e., blocking temperature TB) thermal energy is sufficient to induce free rotation of the particle resulting in a loss of net magnetization in the absence of an external field. This superparamagnetic property, marked by the lack of remnant magnetization after removal of external fields, enables the particles to maintain their colloidal stability and avoid aggregation making it feasible for their use in biomedical applications. It is well known that superparamagnetism is a favourable property of small particles; the reduction of particle size is not without some consequences. They can be labelled as disadvantages for certain applications, specifically as MRI contrast agents. As particle sizes decreases, surface-to-volume ratio increases resulting in pronounced surface effects with negative impact the magnetic properties on the material such as: noncollinear spins, spin canting, and spin-glass-like behaviour, which can significantly negative impact the magnetic properties of the material. Concerning clinical translation, Mankia and coworkers [109] showed that USPIO are already used for limited applications in humans, mainly for the purposes of oncological MRI.

The contrast can arise in MRI from differences in four basic physical parameters:

- ρ: The spin density of the various tissues/fluids being analyzed;
- T_1: The time constant with which the spin magnetization of a given tissue will build up after being saturated/inverted/pulsed-away;
- T_2: The (longer) time constant with which the spins' signals arising from a given tissue will decay away due to homogeneous broadening; this is the kind of irreversible decay that can't be echoed away, arising from microscopic random fluctuations in the magnetic field.
- T_2*: The time constant with which the spin's signals arising from a given tissue will diphase due to inhomogeneous broadening –this is the kind of signal decay that can be echoed away by π pulses; for instance the one arising from field inhomogeneity or susceptibility differences. The corresponding transverse relaxation time constant is thus T_2^* that is usually much smaller than T_2.

Contrast in most MR images is actually a mixture of all these effects; but careful design of the imaging pulse sequence allows one contrast mechanism to be emphasized while the others are minimized. The ability to choose different contrast mechanisms by tailoring the appropriate pulse sequence and choosing the right pulse sequence parameters is what gives MRI its tremendous flexibility.

Malignant and healthy tissues have minor differences in their overall water content. But they exhibit substantial differences in their T_1's & T_2's. Even when this natural contrast is not available MRI distinctions can still be exacerbated via relaxation agents: paramagnetic chemicals that via their isotropic shifts and enhanced relaxation change the T_1, T_2 or T_2* properties of nearby tissues.

Contrast agents are paramagnetic chemicals: chemicals with at least one free, unpaired electron. Unpaired electrons can be found in organic radicals (usually unstable) or in metals (stable). Unfortunately most metals with unpaired electrons are heavy metals: also toxic. To reduce/eliminate the toxicity, metals can be complexated with ligands. This prevents them of interacting biochemically with organs, and makes them wash away quickly. When this does not happen (e.g., when treating renal-compromised patients) there is a risk of a rare but serious illness: nephrogenic systemic fibrosis (NSF). Although a causal link between quelates & NSF has not been definitively established, current guidelines in the United States are that dialysis patients should only receive contrast agents where essential, and that dialysis should be performed as soon as possible after the scan is complete in order to remove the agent from the body promptly.

Typical clinical applications of Gd contrast agents: Vascularity characterization, measure of the healthiness of the blood-brain barrier, tissue angiogenesis &permeability (tumor growth).

Examples of a different, T_2*-based contrast agents, are furnished by superparamagnetic iron oxide (SPIO, USPIO) nanoparticles. This can be used to make MRI contrast agents, which consist of suspended colloids of iron oxide nanoparticles. When injected during an imaging exam they reduce the T_2* signals of absorbing tissues: they yield a hyposignal of the regions where they localize, upon being examined by gradient-echo MRI sequences. SPIOs and USPIOs contrast agents have been used successfully in some instances for liver tumor enhancement.

3.1.3. Extracellular MRI Contrast Agents

The four most commonly extracellular contrast agents approved by the U.S. FDA (Food and Drug Administration) for intravenous use are gadopentetate dimeglumine, gadoteridol, gadodiamide and gadoversetamide [82]. Gd-based extracellular contrast agents are considered safe with a very low overall incidence of adverse events that are mostly minor. Rapid elimination of Gd-chelate from the body is important to minimize the time period of Gd-chelate decomplexation and subsequent toxicity. With all that, in high doses, Gd-based contrast media may be sometimes toxic, particularly in patients with renal insufficiency [137]. Gd-chelates have the theoretical possibility of substantial toxicity from release of the Gd^{3+} ions as well as the free chelator. Free Gd^{3+} ions have a biological half-life of several weeks. Therefore, the ideal chelator must be highly selective and tightly bound to the Gd^{3+} ions to avoid release into circulation. Once released into the body, free Gd^{3+} ions are rapidly sequestered in the bone and liver and may persist for years or even for the patient's lifetime.

Powerful candidates to this application are the magnetite nanoparticles, which exhibit strong T_1 and T_2/T_2^* relaxation properties, significantly larger than those of paramagnetic molecules used as MRI contrast agents such as Gd-chelates [138]. The shortening of the T_2/T_2^* relaxation time associated with accumulation of magnetite nanoparticles in tissues or cells can therefore be used to create a negative contrast on T_2-weighted MRI images, with sensitivity down to the nanomolar range. The ability of magnetite nanoparticles to generate MR contrast is influenced by the particles' size, crystallographic structure and presence of contaminants. The influence of the size and composition of magnetite nanoparticles and other Fe-based spinels on the MR contrast has been studied recently [139, 140]. These studies demonstrated that increasing the size of nanoparticles within the size range of 6 to 12 nm produces an increase in the MR contrasts [138]. This observation agrees with the findings of Goya *et al.* [141] who observed an increase of the magnetization of magnetite nanoparticles with size increase. A higher MR contrast of larger magnetite nanoparticles is reasonable, because faster spin-spin relaxation processes of the water molecules are induced by materials with larger magnetization. However, it has yet to be determined what size range is optimum for highest MR contrast without losing the superparamagnetic properties of nanoparticles. Imaging of the lymph node system has also been achieved through either intravenous or subcutaneous injection of dextrancoated nanoparticles [138]. A major development in the field has been the demonstration of the ability of small lymphotropic superparamagnetic nanoparticles to enable MRI detection of lymph-node metastases in patients with prostate cancer [142].

3.1.4. Cellular MRI, Monitoring Cell Migration for Cell Therapy, Magnetic Resonance Tracking of Dendritic Cells

Using cells as delivery vehicles is one of the most efficiently method in nanotherapeutics. Hence, the need for treating diseases using cellular therapies appears as obvious. Inside the body, the cells can be tracked using optical or bioluminescent imaging, nuclear medicine approaches or magnetic resonance imaging (MRI) techniques.

One of the usefully applications of Cellular MRI is the posibility to develop methods for tracking cells noninvasively inside the body. For this purpose there is need to label cells with superparamagnetic iron oxide nanoparticles (SPION) that have increased NMR relaxation properties (to improve the detection and tracking of therapeutic cells). After that, monitoring the cell migration by magnetic resonance imaging (MRI) techniques [120] should follow.

Joseph A Frank *et al.* [143] have been reported a new strategy to improve the MR properties of labeled stem cells. The method consists in formation of a complex between negatively charged fluorescent monodisperse SPIO nanoparticles and positively charged peptides. This complex formation is used to improve the MR properties of labeled stem cells. Such complex has built in two steps: first, the Fluorescent SPIO nanoparticles (FL SPIO) was synthesized (as described in [144]) by conjugating fixable fluorescent dextran to various SPIO nanoparticles and second, the FL SPIO–protamine sulfate complexes were prepared as previously reported [144, 145] and cells were labeled at optimum complexation conditions for each SPION. The complex solutions of FL SPION–Pro were then used to label the stem cells via the complex formation and compared the labeling efficiency of these fluorescent SPION agents. They also evaluated multimodal nanoparticle-labeled stem cells by fluorescent microscopy and MRI. The developed FL SPION–protamine complex can be used in the future for monitoring stem cell migration for cell therapy.

It is well know that cellular therapies have a powerful dependence on accurate delivery of cells to target organs. In dendritic cell therapy, in particular, delivery and subsequent migration of cells to regional lymph nodes is essential for effective stimulation of the immune system [146]. I Jolanda M de Vries *et al.* showed that *in vivo* magnetic resonance tracking of magnetically labeled cells is feasible in humans for detecting very low numbers of dendritic cells in conjunction with detailed anatomical information. They labeled autologous dendritic cells with a superparamagnetic iron oxide which were injected intranodally in melanoma patients under ultrasound guidance and than, cell migration was tracked by MRI.

Mesenchymal stem cells (MSCs), found in bone marrow, may have enormous therapeutic potential for limiting infarct size and restoring cardiac function after irreversible ischemic injury. Most techniques for the study of stem cell transplantation in animal models require histological analysis to determine the fate and migration of cells wich is rather complicated. Hence, identifying to track stem cells is a real chalange. The recent ability to label MSCs6 with magnetic resonance imaging (MRI)–visible contrast agents should enable serial tracking and quantification of MSC transplantation noninvasively with high spatial resolution. Dara L. Kraitchman proved that magnetically labeled MSCs injected intramyocardially could be detected and tracked noninvasively by MRI in a swine model of myocardial infarction [147]. This noninvasive imaging technique is a promising tool in therapy for myocardial infarction allowing the evaluation of survival, migration, and differentiation status of implanted stem cells in the same subject over time. In their review Rong Zhou *et al.* describes the methods for cell visualization using several corresponding noninvasive imaging modalities, including magnetic resonance imaging, positron emission tomography, single-photon emission computed tomography, and bioluminescent imaging [148].

3.1.5. Gastrointestinal Superparamagnetic Contrast Agents

Gastrointestinal (GI) superparamagnetic contrast agents are used in MRI to improve the visualization of e.g., the stomach, the intestinal tract etc. Suitable materials for oral contrast agents should have little or no absorption by the stomach or intestines, complete excretion, no motion or susceptibility artifacts, affordability, and uniform marking of the gastrointestinal tract. For most body tissues this problem was somehow solved. However, a solution for digestive tract remains a challenge as MRI imaging is very difficult in this case. Literature reports have stated that that only magnetite particles with diameters lower than 10-15 nanometers presents superparamagnetism. However, these particles are absorbed into the bloodstream very quickly. In order not to be taken by the blood stream of the stomach, the use of particles with larger dimensions (micrometric size) could be a suitable solution. The major limitation of micrometric sized particles consists in the difficulty of generating superparamagnetism behaviour. Due to remnant magnetization, the magnetic particles attract each other, making aggregates after magnetic field is removed. An apparent solution to the addressed issue of aggregation was to use, as MRI contrast agents, magnetic microparticles composed by superparamagnetic nanoparticle clusters. In this case clustered nanoparticles present lower saturation magnetization then expected compared to single crystalline particles of the same size.

The solution is developing a synthesis technique able to produce micrometric sized (between 0.5 and 10 micrometers) single crystalline particles (not clusters) having superparamagnetic behaviour (minimizing aggregation process in biologic media). These particles give a combined response to the specific medical application requirements, especially for MRI of the digestive tract as nanoparticulate clusters present relatively significantly higher saturation magnetization (near bulk theoretical one). Beside, by preserving superparamagnetic behavior their larger size make them less for collection in the blood stream (i.e. increasing the bioavailability). The micrometric magnetic particles present weaker reactivity, increased stability against biologic fluids (stomach acidic medium) which is also due to their size.

Next-generation of magnetic particles-based MR imaging contrast agents implies the magnetic core as a major component. Contributions related to new formulations of magnetic particles cores, are welcomed.

Requirements for next generation of magnetic particles-based MR imaging contrast agents should include the following:

1. obtaining biocompatible magnetite, that will be tested in terms of cell viability,
2. obtaning information about optimum size distribution of magnetic particles from cellular endocytosis tests,
3. improving the T_2 signal for obtaining a better contrast in MRI; relaxation time improvement in order to shorter the MRI medical investigation time,
4. avoid the assimilation into the bloodstream of digestive tract,
5. avoid microparticle clusterization after removing the magnetic field (given their superparamagnetic behavior),
6. due to single crystalline structure is expected to have the maximum value for saturation magnetization (close to the bulk),
7. Silva [8] findings suggested that magnetic microparticles once introduced into the stomach are distributed evenly and penetrate the stomach mucosa being stacked there long enough time to allow multiple MRI imaging experiments; the retention time to be increased could be expected.

Based on previous brief and critical literature analysis, there are few things to be summarized:

- Superparamagnetic MRI contrast agents, proved to be potentially a viable alternative in the field;
- The use of superparamagnetic iron oxide based colloids as tissue specific contrast agents is now a well-established area of pharmaceutical development;
- SCMPIO could have a chance to resolve certain unsolved issues in the field of MRI contrast agents, by a sustained research work on magnetic core and coat for contrast agent formulation is needed.

3.2. Drug Delivery Systems

Magnetic micro- and nanoparticle systems play an increasingly important role in medicine, in general, newly applications being rapidly developed, including drug delivery systems (DDS) and technologies.

It is well known that a drug carrier has to meet few essential properties, related both to active pharmacologic carrying agent and to the biologic medium of interaction.

Nanomagnetic particles for biomedical applications must be tuned to the medium so that the formation of large aggregates, changes to their original structure, or biocompatibility are prevented. In particular, magnetic nanoparticles are required to be water-soluble, monodisperse, superparamagnetic at room temperature (no remanence along with rapidly changing magnetic state) [149], and easy to produce on a large scale. However, as has been mentioned the physical and chemical properties of magnetic nanoparticles greatly depend upon the synthesis route [150]. Therefore, the synthesis of nanomagnetite particles that can meet all the above requirements remains a challenge.

Some essentials issues in the drug delivery systems were mentioned in the review chapter, especially those concerning mesoporous and magnetic silica or silica-polymer (PVA) hybrid based matrices and nanocomposites [151, 152].

Drug delivery systems incorporating magnetic materials present complex morphology. First of all, magnetic counterpart itself actually does not always represent a drug vehicle because it is not porous, or because the active agent cannot be directly attached to the magnetic particle surface. For this reason it is necessary to combine the magnetic particle with a protector coating (of inorganic, organic, or biologic nature) in order to create an intermediate shell, which usually was able to linking other functionalities and eventually to bind the active agent. Magnetic particles are used to mainly be a part of more advances DDS, named Controlled Drug Delivery Systems, CDDS. When CDDS are capable of self-control it is about Smart Drug Delivery Systems.

The used method for delivering drugs into systemic circulation plays an important role in the design of NDDSs. In addition to the commonly used oral and injection routes, drugs can also be administered through other means, including oral, parenteral, transdermal, transmucosal, ocular, pulmonary, implantation etc, [153].

The activity of most pharmaceuticals or drugs against certain diseases or disease sites suffers from their inability to gather selectively in the pathological organ or cells. Any kind of drug intravenously administered is distributed in the entire body. This problem may be solved by selectively and quantitatively accumulating the drug to the target site (organ); therefore, magnetic targeting is used when a therapy has limited ability to be chemically targeted to specific types of cells or tissues due to high nonspecific binding. Site-directed drug targeting is one way of local or regional antitumor treatment.

The force felt by a magnetic moment of value $\vec{m}$ in a gradient field is:

$$\vec{F} = (\vec{m} \cdot \vec{\nabla})\vec{B} \tag{1}$$

The magnetic targeting force must compete with the force due to linear blood-flow rates of about 0.05 cm/s in capillaries to 10 cm/s in arteries and 50 cm/s in the aorta. Iron-oxide nanoparticles require flux densities at the target site of the order of 0.1 to 1.0 T with field gradients ranging from 8 T/m (femoral arteries) to over 100 T/m for carotid arteries [85], [154]. The accuracy of magnetic targeting is also dependent on the depth of the target tissue within the body: organs such as the liver and the lungs are harder to target than organs closer to the surface or in the extremities.

One of the difficulties in synthesizing monodisperse iron oxides is the intrinsic agglomeration of magnetic nanoparticles; the formed aggregates are then quickly sequestered by cells of the reticular endothelial system. Other challenges include producing a focused field of sufficient magnitude and gradient and fabricating nanoparticles with sufficiently high moment [68].

To avoid macrophage recognition, particles are surface-protected by a layer of hydrophilic groups. The surface of magnetite nanoparticles can be functionalized with drugs, proteins, and genetic materials to achieve localized delivery of therapeutic agents [155, 156, 157, 158, 159]. Targeted drug delivery by external physical force (magnetic field) is an innovative approach, capable to improve drug targeting. Magnetic drug transport technique is based on the fact that the drug can be either encapsulated into a magnetic micro-sphere (or nano-sphere) or conjugated on the surface of the micro/nano sphere. When the magnetic carrier is intravenously administered, the accumulation can take place within the area to which the magnetic field is applied and often augmented by magnetic agglomeration [160]. The efficiency of accumulation of the magnetic carrier depends on physical and physiological parameters [161], e.g. particle size, surface characteristics, field strength and blood flow rate, etc. The magnetic field assists the movement of the magnetic carrier into a specific tissue. It was observed that immobilization of magnetic target carriers (MTC) occurs within the tumor area when the magnetic field is removed [162]. The efficiency of chemotherapy treatment may be enhanced to a great extent by magnetically assisted delivery of cytotoxic agents to the specific site. Whether an external magnetic field [163] can be used to capture magnetic nanoparticles injected intravenously depends on the strength of the applied magnetic field, as well as on the magnetic properties and volume of the nanoparticles. Alternatively, targeting moieties on the surface of the nanoparticles functionalized with bioactive agents have been used to target tumor in vivo. The recent work from Medarova and colleagues illustrates well the tremendous potential of advanced functional magnetic nanoparticles in the medical field. Simultaneous MR/Near-Infrared optical imaging and tumor delivery of silencing RNA could be achieved using nanoparticles coated with aminodextran and further functionalized with a membrane translocation peptide for intracellular delivery [158]. There are a large number of magnetic carrier systems which demonstrate increasing drug concentration efficiency at the tumor site [164, 165]. In case of brain tumors, the therapeutic ineffectiveness of chemotherapy is mainly due to the impervious nature of the blood-brain barrier (BBB), the presence of drug resistance and lack of tumor selectivity. The BBB is a separation of circulating blood maintained by the choroid plexus in the central nervous system (CNS). Endothelial cells restrict the diffusion of microscopic objects (e.g. bacteria) and large or hydrophilic molecules into the CSF, while allowing the diffusion of small hydrophobic molecules (O_2, hormones, CO_2). Cells of the barrier actively transport metabolic products such as glucose across the barrier with specific proteins.

Various novel biodegradable magnetic drug carriers are synthesized and their targeting to brain tumor is evaluated in vitro and in animal models. New cationic magnetic aminodextran micro spheres (MADM) have been synthesized. Their potential use for drug targeting brain tumor is under investigation [166].

3.3. Gene Delivery

Gene delivery is the process of introducing foreign DNA into host cells. From bacterial to mammalian, for a various types of cells and tissues, many different methods of gene delivery were developed. One of the primary reasons for the limited efficacy of nucleic acid delivery systems is the insufficient contact with target cells. This can be overcome by physical methods of targeting. New chemical approaches in gene therapy allowed for the transfer of genetic information to target cells enabling them to synthesize the protein encoded by the gene [167]. Generally, the methods can be divided into two categories, non-viral and viral. Viral method of gene delivery utilizes the ability of a virus to inject its DNA inside a host cell. Non-viral methods [168], include physical methods such as electroporation, microinjection, gene gun, hydrostatic pressure, continuous infusion, and sonication. Compared to viral ones, nonviral vectors are relatively easy to prepare, less immunogenic and oncogenic, and have no potential of virus recombination and no limitation on the size of a transferred gene. The ability to incorporate genetic materials such as plasmid DNA and RNA, into functionalized nanoparticles demonstrates a new era in pharmacotherapy for delivering genes selectively to tissues and cells [169]. Due to their maneuverability and low toxicity, magnetic particles can be used as a new and effective method of enhancing and targeting nucleic acid delivery [164, 165, 170]. Nucleic acid vectors are associated with magnetic particles and magnetic fields that are used to mediate the rapid contact of these vectors with the target cells. This technique called magnetofection is applicable to viral as well as to non viral vectors. Jin S et al. [171] presented the basic concepts and applications of nonviral gene delivery using super paramagnetic iron oxide nanoparticles and functionalized silica nanoparticles.

The experimental protocols related to these topics are described in their chapter.

3.4. Multimodal Therapy

A multi-modal cancer therapy is a novel concept in nanotherapeutics. This concept involves complementary techniques comprised of magnetic drug targeting followed by simultaneous hyperthermia and drug release. The effects of the therapy should be improved if this process is supervised by MRI techniques.

In order to achieve this, composite nanoparticles consisting of a magnetic iron oxide core and a special thermoresponsive polymeric shell are necessary to be produced. Thus, utilizing the synergistic effect, simultaneous hyperthermia and chemotherapy can be achieved. R V Ramanujan and coworkers at Nanyang Technological University, Singapore, and Singapore General Hospital have developed composite magnetic nanoparticles for this purpose [172].

Shortly, γ- Fe_2O_3 iron oxide magnetic nanoparticles (MNP) with average sizes of 14, 19 and 43 nm were synthesized by high temperature decomposition, followed by coating with the thermoresponsive polymer poly-n-isopropylacrylamide (PNIPAM). *In vitro* and *in vivo* doxorubicin drug release has been tested. The maximum doxorubicin release at 42 °C after 101 h was 41%. *In vitro* simultaneous hyperthermia and drug release of therapeutically relevant quantities of dox was achieved with 14.7% of loaded dox being released in 47 min at hyperthermia temperatures. *In vivo* magnetic targeting of dox-loaded CNP to hepatocellular carcinoma (HCC) in a buffalo rat model was studied by histology and by magnetic resonance imaging. These experiments (doxorubicin-loaded thermoresponsive polymer-coated magnetic nanoparticles) can be a building-block for an efficient tool in multimodal cancer treatment.

3.5. Hyperthermia for Treatment of Cancer, Intracellular Hyperthermia under High-Frequency Alternating Magnetic Field

Temperatures above 42 °C induce cell death in certain tissues. The signs of apoptosis (death of the cell in an orderly manner) begin when cancer cells are heated at temperatures between 41 °C and 47 °C. The increase of the temperature above 50° C is associated to more to necrosis than apoptosis. In order to induce apoptosis, the damages in the mitochondrial membrane are necessary while the mechanism of the synthesis of the proteins remain intact.

Heat treatment of organs or tissues, through increasing the temperature up to 42–46°C is known as hyperthermia [173, 174, 175]. The different hyperthermia techniques depend upon the heating methods used, and each one has certain limitations. Boundary effects limit microwave, ultrasound, and RF hyperthermia. High frequency microwave beams have poor depth penetration and low frequency microwaves are difficult to focus on target areas. Though ultrasound has high penetration and focusing abilities, applications are limited by strong absorption by bone and high reflection by air cavities (lung etc.) [176]. By this technique it is difficult to heat up targets of high perfusion area to the desired temperature due to continuous dissipation of heat. Interstitial technique device is implanted into the tumor, which acts as a heat source if connected to external power sources.

Due to recent research in MRI cell imaging and in cancer magnetic hyperthermia, the interest in living cell labeling using magnetic nanoparticles received more attention. "Magnetic cells" offer exquisite ways for non-contact cell manipulation. Magnetic nanoparticles penetrate the cell, accumulate in endosomes (existing intracellular vesicles) allowing for labelling of various cells by a non-toxic endocytosis mechanism, and place themselves among superparamagnetic iron oxide nanoparticles (SPIONs). Magnetic nanoparticles in a solution can be used as heat sources for magnetic hyperthermia. As described in [177], under the influence of an alternating high-frequency magnetic field, they generate heat through Néel and Brown relaxation processes [178, 179, 180, 181, 182]. For a better understanding of the process a short presentation of the physics of hyperthermia follows.

According to [106], magnetic particles undergo two types of relaxation: Brownian relaxation, in which the entire particle rotates, and Néel relaxation, in which the moment rotates while the particle remains still. The Brownian relaxation time τ_B is:

$$\tau_B = \frac{3\eta V_H}{kT} \qquad (2)$$

where V_H is the hydrodynamic volume of the particle, k is the Boltzmann constant, T is the temperature, and η is the dynamic viscosity of the carrier liquid. The hydrodynamic diameter characterizes how particles move through the fluid where they are suspended. This may be different from the magnetic size due to agglomeration, coating, or interactions between the fluid and the nanoparticle surface. The Néel relaxation time τ_N is defined by:

$$\tau_N = \tau_0 \exp\left(\frac{\Delta E}{kT}\right) \tag{3}$$

where ΔE is the energy barrier over which magnetization must reverse (usually the product of the anisotropy K and the magnetic volume V: ΔE =KV), and τ_0 ~10^{-9}s. Superparamagnetic particles have a relaxation time less than 100 s. The energy barrier depends on the particle size and the relaxation time can accordingly vary 9-10 orders of magnitude. While the Néel relaxation should not depend on the carrier fluid, the Brownian relaxation time is highly dependent on fluid viscosity, which can be controlled. The two relaxation processes are also differently impacted by the immobilization of nanoparticles due to endocytosis.

The first reported use of magnetite nanoparticles in hyperthermia was by Gilchrist *et al.* in 1960 [183], and since then there has been tremendous progress in this field. Later this procedure was achieved by direct injection of nanoparticles into the tumors. If sufficient quantity of the magnetite nanoparticles injected into the tumor to maintain a temperature of more than 42 °C for 30 min, tumor cells could be destroyed and adjacent healthy cells remain unaffected. The nanoparticles can be introduced into the body and then magnetically targeted to the tumor, or injected directly into the tumor, or injected into the vasculature supplying the tumor.

The first clinical human trials using magnetic hyperthermia were reported by Lübbe *et al.* [166], [76], [184] who used 100-nm starch-coated iron-oxide particles bound with epirubicin for treatment of advanced solid cancers. Positive results from ongoing trials of advanced cancer patients who received magnetic nanoparticle hyperthermia in conjunction with conformal external beam radiation therapy were recently reported by Jordan [177]. The therapy was well tolerated by the patients meanwhile significant increases in the length and quality of life were observed [84]. Intracellular hyperthermia is a technique where fluids containing submicron-sized magnetic particles (typically 1–100 nm) are injected. These particles are easily incorporated into cells, since their diameters are in the nanometer range. These magnetic particles selectively heat up tissues by coupling AC magnetic field to targeted magnetic nanoparticles. As a result, the whole tumor can be heated up uniformly. The technique is based on the fact that tumor cells are more sensitive to temperature than normal cells, and they have a very good appetite for iron oxide (nine times more than normal cells).

Endocytosis of these magnetic nano-heaters localizes the source of heat in the internal volume of the cell, at the endosome level. It demonstrates that intracellular confinement of nanoparticles abrogates the Brownian contribution to heating. In other words, intracellular hyperthermia mostly originates from Néel relaxation [185]. The resulting temperature elevation is enough to assure tumor cell killing efficiency of intracellular hyperthermia [186, 187, 188]. The main challenge now lies in selective intracellular delivery to cancer cells in vivo, as this would enable tumor cells to be killed without damaging normal by-stander cells and tissues. To achieve sufficient heat elevation, it becomes necessary to examine the best nanoparticle design for such heating applications. In particular, increasing maghemite nanoparticles' size to about 12–14 nm in diameter, with low polydispersity index (0.1 – 0.2) should increase the heating power at least tenfold [76].

Considerable interest has therefore been drawn to magnetic nanoparticles to achieve intracellular hyperthermia procedures, for instance by targeting magnetic nanoparticles conjugated with a monoclonal antibody such as anti-HER2 [137]. Magnetic nanoparticles exposed to an alternating magnetic field can indeed generate heat by hysteresis loss (multidomain particles) or energy dissipation (superparamagnetic nanoparticles). Proof-of-principle has been demonstrated by a few groups in vivo [187], [189, 190, 191, 192]. In addition, the synergistic combination of hyperthermia procedure and local delivery of chemotherapeutics holds promises of improved therapeutic activity. As for other in vivo diagnostic and therapeutic applications, the major challenge is to accumulate with high specificity the magnetic nanoparticles in the target tissues.

Claire Wilhelm and Florence Gazeau [178] described a simple and efficient method for labelling nanoparticles made of maghemite or cobalt ferrite and chelated with citrate ions; they were dispersed in cell culture medium supplemented with 5mM citrate at an iron concentration of 5mM. They demonstrated that conferring magnetism inside the cells opens opportunities for mechanical manipulations to therapeutic applications. Magnetic nanoparticles are enclosed inside intracellular vesicles that can be deformed by applying a permanent magnetic field to probe intracellular membrane or to be rotated by applying a pulse rotation or an oscillating magnetic field to probe their intimate environment. The multifunctional magnetic endosomes should open new issues for the mechanical manipulations of intracellular organelles. In particular, one could combine intracellular heating and intracellular microrheology to explore the mechanical damages provoked by local heating as well as cytoplasm repair after heating. In that case, nanoparticles would be successively heaters and probes.

Amit P. Khandhar et al. have developed a comprehensive protocol for synthesizing highly monodispersed MNPs in organic solvents, subsequently transferred to water using a biocompatible amphiphilic polymer. These MNPs show truly size-dependent heating rates, indicated by a sharp peak in the specific loss power (SLP, W/g Fe_3O_4) for 16 nm (diameter) particles. They performed heating measurements in biological medium (fetal bovine serum) and concluded that Brownian relaxation contribution is blocked when particles size increases and agglomeration appears. Finally, they concluded that for optimization of Magnetic Fluid Hyperthermia it is preferable to use monodispersed MNPs that relax predominantly via Néel mechanism, which is unaltered by changes in hydrodynamic volume that are inevitable for foreign objects circulating in vivo [193]. A combinatory effects against cancer cells, was analyzed by Hojjat Sadeghi-Aliabadi *et al.* Their results showed that co-administration of doxorubicin and magnetite nanoparticles in an alternate (AC) magnetic field could be a cytotoxic remedy on breast cancer cells via hyperthermia. The nanoparticles used in these experiments (3-10 nm) were obtained by co-precipitation and stabilized by oleic acid [194]. Magnetite nanoparticles loaded with doxorubicin and tetrandrine were used against K562 leukemia cells as a new strategy to inhibit drug resistance [195]. Water dispersible Fe_3O_4 nanoparticles carrying doxorubicin [196] or docetaxel [197] have been prepared for cancer therapy.

Cheraghipour et al. synthesized by co-precipitation magnetite nanoparticles of about 10 nm. The particles were capped with citric acid (reactive carboxyl groups) on their surface. Citrate capped MNPs had remarkable heating effect during the application of a magnetic field, which make them attractive for biomedical heating applications, such as magnetic fluid hyperthermia [198]. Currently, magnetite and maghemite are the only FDA approved magnetic materials that can be used in humans [193].

3.6. Cell Labeling, Cell Separation

3.6.1. Immunomagnetic Separation of Cells, Detection, Immobilization and Modification of Biologically Active Compounds, Cell Labeling

In the last years, maghemite or magnetite nanoparticles have shown great attractive applications including cellular labeling and separation. Due to their stability and biocompatibility, many researchers have focused attention on the separation, detection and target capability of maghemite nanoparticles in earlier diagnosis and treatment for some diseases. Conjugated with biomolecules such as antibodies, biotin and enzyme and others, the MNPs could form lock-and-key interactions with target molecules of antigenNPs offering a powerful instrument for specific recognition and target capability. Therefore, is very important to investigate the perspectives of conjugation efficiency, conjugation activity and the conjugation stability of antibodies on the surface of MNPs. In this respect, the synthesis of the core/shell magnetic nanoclusters and the attachment of active ligands to their surface are very important operations for applications in bio-medicine. Coating can be performed with organic shells, or with inorganic materials such as gold, silica, or carbon.

Invertase was immobilized by various methods: by immersion/dipping the support in enzymatic solution, by direct attachment via carbodiimide activation and by activation with sodium citrate to magnetic nanoparticles support [199]. The magnetic nanometric support was synthesized by thermal co/precipitation of ferric and ferrous chloride ions using different work conditions; 4 support types were obtained. The supports were characterized by Fourier Transform Infrared (FTIR). The efficiency of the enzyme immobilization on different supports was calculated. Biologically active compounds [200, 201] immobilized on magnetic carriers can be removed from the system by using an external magnetic field or can be targeted to the desired place. They can be used to express their activities in the process or as affinity ligands enabling the capture or modification of the targeted molecules or cells. This technology can be used to modify antibodies. Magnetically-labeled antibodies have been proposed for clinical applications as a therapeutic agent for the induction of hyperthermia.

A very efficient stem cell labeling with coated gamma-Fe_2O_3 nanoparticles was reported by Eva Syková *et al.* [202]. In this study iron oxide cores were prepared through the well-known precipitation of iron salts, in particular $FeCl_2$ and Fe Cl_3, by increasing pH with ammonium hydroxide, followed by the oxidation of the resulting magnetite with sodium hypochlorite. Oxidation resulted in maghemite, which is chemically stabile and does not change its properties. This is in contrast to maghemite, which undergoes uncontrolled oxidation in the presence of oxygen.

Another example can be the specific anti-human cTnI antibody conjugated with MNPs (immunomagnetic nanoparticles, IMNPs), separated by a gradient magnetic field. The cardiac troponin I (cTnI) is a subunit of cardiac troponin complex that is broken up in the individual protein components which are released into the bloodstream when the myocardial damage occurs. Therefore, cTnI in the serum of patients has been considered as the best biomarker for diagnosis of myocardial damages. For this purpose, Shang *et al.* [203] focused on the preparation of the IMNPs, and the conjugation amount of antibodies. The biological activity and the stability of the IMNPs by biological method of the enzyme linked immunosorbent assay (ELISA) and Western blotting has been studied and for the first time; the morphology and the stability of IMNPs resulted. Dimercaptosuccinnic acid (DMSA)-coated MNPs were synthesized by chemical coprecipitation and characterized by TEM, XRD, Zeta Potential Analyzer and VSM. The anti-human cTnI IMNPs were prepared by grafting antibodies on the surface of DMSA-coated MNPs using EDC cross-linker. The conclusion shows that the antibodies remain their specific affinity and biological stability after 300 days. The maximum conjugation amount of antibodies on the 0.1 mg MNPs is about 96 μg by covalent bonds.

In [204] a simple strategy was presented for tailoring the surface of superparamagnetic iron oxides nanocrystals using a multifunctional polymeric ligand resulting in a phase transfer from nonpolar to polar solvents, at the same time preventing their agglomeration. The copolymer functionalized nanoparticles are used for immobilizing and subsequently isolating His-tagged recombinant protein silicate in marine sponge suberites domuncula. This highly versatile biomagnetic separation method allows not only the re-use of the magnetic nanocrystals but makes it a simple and faster method for purification and effective manipulation of various kinds of engineered proteins as an alternative to existing protocols for protein separation.

3.6.2. Magnetic Separation of Cells and Bio Molecule; Bio-Assays

The detection of specific molecules is critical for diagnosis, treatment, and prevention of disease. A salient feature of superparamagnetic materials is the fact that they exhibit magnetic properties only in the presence of a magnetic field. This makes them very effective in the process of magnetic separation of cells or bio molecules. Magnetic separation processes have been widely investigated and used in many biotechnology applications such as purification, cell sorting, and immunoassays [138]. Magnetic nanoparticle-based separation processes are attractive as they can be efficiently operated with small quantity of target entities (e.g., protein markers). Magnetic cell sorting was first proposed by using surface markers for cell receptors [205]. Magnetic cell separation allows for the separation of target cells directly from blood, bone marrow and other fluids in short times due to fast reaction kinetics.

These techniques are based on the contrast of magnetic susceptibility between medium (containing other nonmagnetic) and separand (magnetic). The process is also clearly described by D. Bahadur and Jyotsnendu Giri [201].

Immobilized protein A or protein G on silanized magnetite and fine magnetotactic bacteria can be used for isolation and purification of IgG. Monosized superparamagnetic particles, Dynabeads, have been used in isolation of mRNA, genomic DNA and proteins. Isolation and separation of cells (prokaryotic, eukaryotic) or antigen have been done by immobilization of specific antibodies against the target (cells or antigen) i.e. by immunomagnetic separation [67], [206, 207]. Magnetic nanoparticles represent a very good alternative to conventional biolabels, which rely on fluorescence and chemiluminescence [208, 209, 210]. Using ferromagnetic "spin valve" sensors, very sensitive magnetic microarrays are for instance being developed by Wang and co-workers to detect binding of target DNA and proteins [160]. The electrical resistance of these sensors changes in a predictable way in the presence of single domain magnetic nanoparticles captured on the surface of the chip in a sandwich assay strategy [210].

Removal of cancer cells from bone marrow is one of the most important applications of magnetic separation. Tumor cell separation from peripheral blood has been performed by immobilization of antibody on silica coated with superparamagnetic iron oxide [211].

The use of magnetic nanoparticles in bio-assay is strongly correlated with the optimization of molecular recognition processes. Lin *et al.* recently reported the development of ethylene glycol protected magnetite nanoparticles for a Matrix-Assisted Laser Desorption/Ionization Mass Spectroscopy (MALDI-MS)–based immunoassay [212]. The presence of ethylene glycol groups [139] on nanoparticles significantly reduced non-specific binding and consequently substantially improved the sensitivity of the proposed immunoassay. In clinical application, magnetite nanoparticles coated with the monoclonal antibody towards the human Epithelial Cell Adhesion Molecule (EpCam) have been used to specifically enrich epithelial tumor cells from biological samples (e.g., whole blood or bone marrow) [213].

The high specificity of biomolecular interactions, such as antibody/antigen, avidin/biotin interactions, is essential for the immune system to provide the organism with the ability to recognize, respond to, and in most cases successfully defend against a wide variety of infection agents. Given the potential chemical, biological and clinic importance of this high affinity selective observation, significant efforts have been made to investigate the behaviors of molecular recognition.

Multifunctional polymer-derivatized superparamagnetic iron oxide (gamma-Fe_2O_3) nanoparticles were prepared for biomagnetic separation of histidine-tagged recombinant proteins building up a faster and efficient method for protein separation by making use of their intrinsic magnetic properties [182]. Using polymer bound gamma-Fe_2O_3 nanocrystals, a 6-histidine-tagged recombinant protein (silicatein) with a molecular weight of 24 kDa has been isolated and purified. The supermagnetic iron oxide nanocrystals were characterized by transmission electron microscopy (TEM), high-resolution TEM (HRTEM), SQUID and Mössbauer and the polymer functionalization of the gamma-Fe_2O_3 nanocrystals was monitored by UV–vis spectroscopy and light microscopy. Protein immobilization and separation were monitored using immunostaining techniques and gel electrophoresis, respectively.

In [214], the superparamagnetic magnetite/dextran/IgG nanocomposites were used as immunolabels to specifically recognize goat anti-mouse immunoglobulin (anti-IgG), which was covalently attached onto a gold substrate. Such magnetic immunolabels have recently attracted broad scientific attention, both fundamental and applied, since they possess high magnetic susceptibility, low remanence, low coercivity and high saturation magnetization, and have been widely used in bioseparation, targeted drug delivery, immunoassays and biosensors. The recognition processes were detected by AFM, MFM and infrared reflection spectra. Moreover, MFM was operated to distinguish impurities from the immunorecognized nanocomposites, which would be helpful for bioassay and biosensor. In [215] a step-by-step approach to prepared immunological nanomagnetic films was described. Magnetic-labeled IgG was used as an immunospecific reagent to recognize anti-IgG surface. AFM and infrared spectroscopy verified the success of each immobilization step. Detecting the magnetic signal by MFM allowed obtaining the information of the specific recognition. Roughness, aggregation and absorption of impurities could not affect the final detection result. MFM was found to be more accurate than AFM. Very recently [216], it was reported the incorporation of magnetic gamma-Fe_2O_3 nanoparticles of 6 nm diameter into polyorganosiloxane nanospheres allowing different functionalities in the future. The big advantage is the potential of modified surface of the polyorganosiloxane (for instance) with poly (ethylene oxide) to produce nanospheres water dispersible, also allowing, for example, cell uptake experiments. This study better enhanced the understanding of the biomolecular recognition process and it has great promise for the biomedical, bioengineering, diagnostic and therapeutic applications.

CONCLUSION

Survey of the literature shows that many new applications are based on the use of iron oxide particles with wide variety of sizes, coatings, and chemical specific destinations; it is difficult and impractical to present them on all here. For this reason, this chapter was not intended to be anything more than a general overview in the field.

This chapter outlines the main properties of iron oxide nanoparticles and their various bio-medical applications. A brief presentation is provided in which novel approaches to medical applications are described. The essential SPIO-concept and its derivate are presented. Modern applications were discussed in cell separation, as contrast agents in magnetic resonance imaging, drug and gene delivery, hyperthermia applications, as well as the very new concepts in nanotherapeutics: the cell therapy and multimodal cancer therapy. We underlined the efforts to extend the superparamagnetic behavior from nanometric level to micrometric level; we showed that extending this advantage from nanometric to micrometric scale is a real challenge and could be very useful, in the biomedical field. We underlined the specific properties of magnetite with micrometer dimensions and superparamagnetic behavior for a single crystalline structure. The number of applications of micromaterials to photoelectrochemistry and biotechnology/biosciences are gradually increasing, and this represents a challenging area for future research. Interdisciplinary research includes physics, chemistry, biology & medicine, engineering and others sciences to try shed light on new routes for healthy applications. The enormous potential of magnetic nano/microparticles applications continues to stimulate and enhance our imagination. Micromaterials were proven to contribute to the solutions of many critical current problems, opening new highways in interdisciplinary sciences.

ACKNOWLEDGMENTS

The authors would like to thank Dr. Adrian Ieta from Department of Physics, State University of New York at Oswego, Oswego, NY for reading the manuscript. This work was supported by the project POSDRU ID 137070 and PN 09-340107 of the Ministry of Research and Education, Romania.

REFERENCES

[1] A. Y. I. Al-Baitai, Computational Studies of the Interaction of Pollutants with Iron Oxide Surfaces, London: University College London, 2011.

[2] M. Chirita and I. Grozescu, „Iron Oxide Nanoparticles and Their Photoelectrochemical and Medical Applications A Short Overview," *Chem. Bull.* "Politehnica" Univ.Timisoara, vol. 54, nr. 68, p. 1, 2009.

[3] R. M. Cornel and U. Schwertmann, *The Iron Oxides. Structure, Properties, Reactions and Uses,* VCH:Weinheim, 1996.

[4] V. Runge, "Contrast Media in: Findings advise against lower MRI contrast doses," *Diagnostic Imaging,* p. 23–24, 1998.

[5] K. S. Mankia, M. A. McAteer and R. P. Choudhury, "Microparticle-Based Molecular MRI of Atherosclerosis, Thrombosis, and Tissue Ischemia,," *Curr Cardiovasc Imaging Rep.,* vol. 4, p. 17–23, 2011.

[6] M. Carroll, R. Woodward and. M. J. House, "Experimental validation of proton transverse relaxivity models for superparamagnetic nanoparticle MRI contrast agents," *Nanotechnology,* vol. 21, no. 3, 2010.

[7] I. Y. Chen and J. C. Wu, "Cardiovascular Molecular Imaging: Focus on Clinical Translation Circulation," *Circulation,* vol. 123, no. 4, pp. 445-443, 2011.

[8] É. L. Silva, J. F. Carvalho, T. Pontes, E. E. Oliveira, B. L. Francelino, A. Medeiros, E. Egito, J. Araujo and A. Carriço, "Development of a magnetic system for the treatment of Helicobacter pylori infections," *Journal of Magnetism and Magnetic Materials,* vol. 321, no. 10, p. 1566–1570, 2009.

[9] C. von zur Muhlen, K. Peter, Z. Ali, J. E. Schneider and M. A. McAteer, "Visualisation of activated platelets by targeted magnetic resonance imaging utilizing conformation-specific antibodies against glycoprotein IIb/IIIa," *J. Vasc. Res.,* vol. 46, no. 1, pp. 6-14, Neubauer S, Channon KM, Bode C, Choudhury RP, 2009.

[10] A. Akhtar, J. Schneider and. S. Chapman, "In vivo quantification of VCAM-1 expression in renal ischaemia reperfusion injury using non-invasive magnetic resonance molecular imaging," *PLoS ONE,* vol. 5, no. 9, p. 12800, 2010.

[11] M. A. McAteer, C. von Zur Muhlen, D. C. Anthony, N. R. Sibson and P. Robin, "Choudhury Magnetic Resonance Imaging of Brain Inflammation Using Microparticles of Iron Oxide Methods in Molecular Biology," pp. 103-115, 2011.

[12] M. Jean-Baptiste and. M. Sylvain, "Magnetic Steering of Iron Oxide microparticles Using Propulsion Gradient Coils in MRI," New York City, USA, 2006.

[13] L. Liu, H.-Z. Kou, W. Mo, H. Liu and Y. Wang, "Surfactant-Assisted Synthesis of α-Fe2O3 Nanotubes and Nanorods with Shape-Dependent Magnetic Properties," *J. Phys. Chem. B,* vol. 110, no. 31, p. 15218–15223, 2006.

[14] R. Chen, J. Cheng and Y. Wei, "Preparation and magnetic properties of Fe3O4 microparticles with adjustable size and morphology," *Journal of Alloys and Compounds 520,* vol. 520, p. 266– 271, 2012.

[15] H. Qi, J. Ye, N. Tao, M. Wen and Q. Chen, "Synthesis of octahedral magnetite microcrystals with high crystallinity and low coercive field," *Journal of Crystal Growth,* vol. 311, no. 2, pp. 394-398, 2009.

[16] M. Hu, R. P. Ji and J. S. Jiang, "Hydrothermal synthesis of magnetite crystals: From sheet to pseudo-octahedron," *Materials Research Bulletin,* vol. 45, no. 12, p. 1811–1815, 2010.

[17] H. Deng, H., X. Li, Q. Peng, X. Wang, J. Chen and Y. Li, "Monodisperse Magnetic Sigle - Crystal Ferrite Microspheres," *Angew. Chem.,* vol. 117, no. 18, p. 2842–2845, 2005.

[18] M. Chirita, R. Banica, P. Sfarloaga, A. Ieta and I. Grozescu, "A Short Route of Micrometric Magnetite Synthesis via Fe-EDTA Thermal Decomposition," in International Semiconductor Conference IEEE Conf Proc Sinaia, Sinaia, 2010.

[19] M. Chirita, R. Banica, A. Ieta and I. Grozescu, "Superparamagnetic Unusual Behavior of Micrometric Magnetite Monodisperse Monocrystals Synthesized by Fe-EDTA Thermal Decomposition," *Particulate Science and Technologies,* vol. 30, no. 4, pp. 354-363, 2012.

[20] M. Chirita and A. Ieta, "FeCO3 Microparticle Synthesis by Fe-EDTA Hydrothermal Decomposition," *Crystal Growth and Design,* vol. 12, pp. 883-886, 2012.

[21] G. Hägg, "Solid Solutions with a Varying Number of Atoms in the Unit Cell," *Zeitschrift für Kristallographie-Crystalline Materials,* vol. 91, no. 1, p. 114–122, 1935.

[22] P. Krishna and D. Pandey, Close-Packed Structures, Cardiff.: Printed in Wales by University College, 2001.

[23] R. J. Hill, J. R. Craig and G. V. Gibbs, "Systematics of the spinel structure type," *Physics and Chemistry of Minerals,* vol. 4, pp. 317-340, 1979.

[24] A. Kay, I. Cesar and M. Gratze, "lNew benchmark for water photooxidation by nanostructured α- Fe2O3 films," *Journal of Yongqi Liang* et al. *the American Chemical Society, vol. 128, no. 49,* p. 15714–15721, 2006.

[25] Y. Qiang, A. Sharma, J. Nutting, D. Sikes and D. Meyer, "Iron/iron oxide core-shell nanoclusters for biomedical applications," *Journal of nanoparticle Research,* vol. 8, no. 3-4, pp. 489-496, 2006.

[26] C. Pascal, J. L. Pascal, F. Favier, M. Elidrissi Moubtassim and C. Payen, "Electrochemical Synthesis for the Control of γ-Fe2O3 Nanoparticle Size. Morphology, Microstructure, and Magnetic Behavior," *Chem. Mater.,* vol. 11, no. 1, pp. 141-147, 1999.

[27] G. Ennas, A. Musinu, G. Piccaluga, D. Zedda, D. Gatteschi, C. Sangregorio, J. Stanger, G. Concas şi G. Spano, „Caracterization of iron oxide nanoparticles in an Fe2O3-SiO2composite prepared by a sol-gel method," *Chem. Mater,* vol. 10, pp. 495-502, 1998.

[28] S. Mørup, F. Bodker, P. V. Hendriksen and S. Linderoth, "Spin-glass-like ordering of the magnetic moments of interacting nanosized maghemite particles," *Phys. Rev. B,* vol. 52, p. 287, 1995.

[29] E. Tronc, J. P. Jolivet and J. Livage, "Mössbauer investigation of the γ→α-Fe2O3 transformation in small particles," *Hyperfine Interact.,* vol. 54, pp. 737-740, 1990.

[30] E. Scolan and C. Sanchez, "Synthesis and characterization of surface-protected nanocrystalline titania particles," *Chem Mater 10,* p. 3217–3223, 1998.

[31] C. Savii, M. Popovici, C. Enache, J. Subrt, D. Niznansky, S. Bakardzieva, C. Caizer and I. Hrianca, "Fe2O3-SiO2 composites obtained by sol-gel synthesis," *Solid State Ionics,* vol. 151, no. 1-4, pp. 219-227, 2002.

[32] M. Popovici, M. Gich, M. Roig, L. Casas, E. Molins, C. Savii, D. Becherescu, J. Sort, S. Surinach, J. S. Munoz, M. D. Baro and J. Nogues, "Ultraporous single phase iron oxide-silica nanostructured aerogels from ferrous precursors," *Langmuir,* vol. 20, no. 4, pp. 1425-1429, 2004.

[33] N. Raschzok, C. M. Langer, C. Schmidt, K. H. Lerche, N. Billecke, K. Nehls, N. B. Schlüter, A. Leder, S. Rohn, M. T. Mogl, L. Lüdemann L, L. Stelter, U. K. Teichgräber, P. Neuhaus and I. M. Sauer, "Functionalizable silica-based micron-sized iron oxide particles for

cellular magnetic reso," *Cell Transplant,* vol. 22, no. 11, pp. 1959-1970, 2013.

[34] P. Linlin, X. Tengfeng, F. Zhiyong, Z. Qidong, W. Dejun and Z. Dan, "Surface photovoltage characterization of an oriented a- Fe2O3 nanorod array," *Chemical Physics Letters 459,* p. 159–163, 2008.

[35] Y. Fu, R. Wang, J. Xu, J. Chen, Y. Yan, A. Narlikar and H. Zhang, "Synthesis of large arrays of aligned a-Fe2O3 nanowires," *Chem. Phys. Lett. 379,* p. 373, 2003.

[36] N. Beermann, L. Vayssieres, S. Lindquist and A. Hagfeldt, "Photoelectrochemical Studies of Oriented Nanorod Thin Films of Hematite," *J. Electrochem. Soc. 147,,* p. 2456, 2000.

[37] S. Zhan, D. Chen, X. Jiao and S. Liu, "Facile fabrication of long alpha-Fe2O3, alpha-Fe and gamma-Fe2O3 hollow fibers using sol–gel combined co-elecrospinning technology," *J. Colloid Interface Sci., 308,* p. 265–270, 2007.

[38] C. Wu, P. Yin, X. Zhu, C. Yang and Y. Xie, "Synthesis of hematite (alpha-Fe2O3) nanorods: diameter-size and shape effect on their applications in magnetism, lithium ion battery, and gas sensors," *J. Phys. Chem. B,* vol. 110, pp. 17806-17812, 2006.

[39] C. Caizer, M. Popovici and C. Savii, "Spherical (Zn delta Ni1-delta Fe2O4)(gamma) nanoparticles in an amorphous (SiO2)(1-gamma) matrix, prepared with the sol-gel method," *Acta Mater.,* vol. 51, no. 12, pp. 3607-3616, 2003.

[40] C. Caizer, C. Savii and M. Popovici, "Magnetic behaviour of iron oxide nanoparticles dispersed in a silica matrix," Materials Science And Engineering B-Solid State Materials For Advanced Technology, vol. 97, no. 2, pp. 129-134, 2003.

[41] Q. Liu, Z. Cui, Z. Ma, S. Bian, W. Song and L. J. Wan, "Morphology control of Fe2O3 nanocrystals and their application in catalysis," *Nanotechnology,* vol. 18, p. 385605, 2007.

[42] L. Vayssieres, C. Sathe, S. Butorin, D. Shuh, J. Nordgren and J. H. Guo, "1-D Quantum Confinement Effect in a-Fe2O3 Nanorod Arrays," *Adv. Mater.,* vol. 17, pp. 2320-2323, 2005.

[43] K. Cheng, Y. He, Y. Miao and e. al, "K. Cheng, YP He, YM Miao et al. J. Phys. Chem. B., 110 (2006), p. 7259," *J. Phys. Chem. B 110,* p. 7259, 2006.

[44] G. Mor, H. Prakasam, O. Varghese, K. Shankar and C. A. Grimes, "Vertically oriented Ti-Fe-O nanotube array films: toward a useful

material architecture for solar spectrum water photoelectrolysis," *Nano Lett.,* vol. 7, no. 8, pp. 2356-2364, 2007.

[45] J. Wang, X. Sun, Y. Yang, H. Huang and Y. Lee, "Hydrothermally grown oriented ZnO nanorod arrays for gas sensing applications," *Nanotechnology,* vol. 17, no. 19, p. 4995, 2006.

[46] Y. Li, B. Tan and Y. Wu, "Freestanding mesoporous quasi-single-crystalline Co3O4 nanowire arrays," *J. Am. Chem. Soc.,* vol. 128, no. 44, pp. 14258-14259, 2006.

[47] L. Vayssieres and M. Graetzel, "Highly ordered SnO2 nanorod arrays from controlled aqueous growth," *Angew. Chem. Int. Ed.,* vol. 43, p. 3666, 2004.

[48] M. Paulose, K. Shankar, S. Yoriya, H. E. Prakasam, O. K. Varghese, G. K. Mor, T. A. Latempa, A. Fitzgerald and C. A. Grimes, "Anodic Growth of Highly Ordered TiO2 Nanotube Arrays to 134 μm in Length," *J. Phys. Chem. B,* vol. 110, no. 33, pp. 16179-16184, 2006.

[49] D. Routkevitch, T. Bigioni, M. Moskovits and J. Xu, "Electrochemical fabrication of CdS nanowire arrays in porous anodic aluminum oxide templates," *J. Phys. Chem.,* vol. 100, no. 33, pp. 14037-14047, 1996.

[50] X. Wen, S. Wang, Y. Ding, Z. Wang and S. Yang, "Controlled growth of large-area, uniform, vertically aligned arrays of a-Fe2O3 nanobelts and nanowires," *J. Phys. Chem. B,* vol. 109, pp. 215-220, 2005.

[51] T. Lindgren, H. L. Wang, N. Beermann, L. Vayssieres, A. Hagfeldt and S. E. Lindquist, "Anqueos photoelecrochemistry of hematite nanorods array," *Sol. Energ. Mater. Sol. C,* vol. 71, no. 2, pp. 231-243, 2002.

[52] V. Satsangi, S. Kumari, A. Singh, R. Shrivastav and S. Dass, "Nanostructured hematite for photoelectrochemical generation of hydrogen," *Int. J. Hydrogen Energ.,* vol. 33, no. 1, pp. 312-318, 2008.

[53] A. Tavakoli, M. Sohrabi and A. Kargari, "Review of Methods for Synthesis of Nanostructured Metals with Emphasis on Iron Compounds," *Chem. Pap.,* vol. 61, no. 3, pp. 151-170, 2007.

[54] R. Lane, B. Craig and W. Babcock, "Materials engineering with nature's building blocks," *AMPTIAC Newslett Spring,* vol. 6, no. 1, p. 31–37, 2002.

[55] K. J. Klabunde, Nanoscale Materials in Chemistry Chapter 4., New York: Wiley, 2001.

[56] S. C. Tjong and H. Chen, "Nanocrystalline materials and coatings," *Mater. Sci. Eng,* vol. 45, no. 1, pp. 1-88, 2004.

[57] R. Herrera-Becerra, C. Zorrilla., J. L. Rius and J. A. Ascencio, "Electron microscopy characterization of biosynthesized iron oxide nanoparticles," *Appl. Phys. A,* vol. 91, pp. 241-246, 2008.

[58] S. Qu, H. Yang, D. Ren, S. Kan, D. Zou, M. Li and J. Li, "Magnetite Nanoparticles Prepared by Precipitation from Partially Reduced Ferric Chloride Aqueous Solutions," *J. Colloid Interf. Sci.,* vol. 215, no. 1, pp. 190-192, 1999.

[59] R. Massart, " Preparation of aqueous magnetic liquids in alkaline and acidic media," *IEEE Trans. Magn.,* vol. 17, no. 25, pp. 1247-1248, 1981.

[60] R. Massart and V. Cabuil, "Effect of some parameters on the formation of colloidal magnetite in alkaline-medium-yield and particle-size control," *J. Chem. Phys,* vol. 84, p. 967–973, 1987.

[61] T. Osaka, T. Matsunaga, T. Nakanishi, A. Arkaki, D. Niwa and H. Iida, *Anal. Bioanal. Chem. 384, 593,* vol. 384, pp. 593-600, 2006.

[62] S. Sun, H. Zeng, D. B. Robinson, S. Raoux, P. M. Rice, S. X. Wang and G. Li, "Monodisperse MFe2O4 (M = Fe, Co, Mn) nanoparticles," *J. Am. Chem. Soc. 2004, 126, 273,* vol. 126, no. 1, pp. 273-279, 2004.

[63] S. Peng, C. Wang, J. Xie and S. Sun, "Synthesis and stabilization of monodisperse Fe nanoparticles," *J. Amer. Chem. Soc.,* vol. 128, p. 10676–10677, 2006.

[64] K. Woo, J. Hong, S. Choi, H. W. Lee, J. P. Ahn, C. S. Kim and S. W. Lee, "Easy Synthesis and Magnetic Properties of Iron Oxide Nanoparticles," *Chem Mater,* vol. 16, no. 14, pp. 2814-2818, 2004.

[65] A. L. Andrade, M. A. Valente, M. F. Ferreira and J. D. Fabris, "Preparation of size-controlled nanoparticles of magnetite," *Journal of Magnetism and Magnetic Materials,* vol. 324, p. 1753–1757, 2012.

[66] J. W. Bulte, T. Douglas, B. Witwer, S. C. Zhang, E. Strable, B. K. Lewis, H. Zywicke, B. Miller, P. van Gelderen, B. M. Moskowitz, I. D. Duncan and J. A. Frank JA., "Magnetodendrimers allow endosomal magnetic labeling and in vivo tracking of stem cells," *Nat Biotechnol.,* vol. 19, no. 12, pp. 1141-1147, 2001.

[67] O. Olsvik, T. Popovic, E. Skjerve, K. S. Cudjoe, E. Hornes, J. Ugelstad and M. Uhlén, "Magnetic separation techniques in diagnostic microbiology," *Clin. Microbiol. Rev.,* vol. 7, no. 1, pp. 43-54, 1994.

[68] C. Alexiou, W. Arnold, R. J. Klein, F. G. Parak, P. Huh, C. Bergemann, W. Erhardt, S. Wagenpfeil and A. S. Lubbe, "Locoregional cancer treatment with magnetic drug targeting," *Cancer Res.,* vol. 60, pp. 6641-6648, 2000.

[69] N. Kohler, C. Sun, A. Fichtenholtz, J. Gunn and C. Z. Fang, "Methotrexate-immobilized poly(ethylene glycol) magnetic nanoparticles for MR imaging and drug delivery," *Small,* vol. 2, no. 6, pp. 785-792, 2006.

[70] Y. Yang, J. S. Jiang, B. Du, Z. F. Gan, M. Qian and P. Zhang, "Preparation and properties of a novel drug delivery system with both magnetic and biomolecular targeting," *J. Mater Sci: Mater Med.,* vol. 20, no. 1, pp. 301-307, 2009.

[71] T. K. Jain, M. A. Morales, S. K. Sahoo, D. L. Leslie-Pelecky and V. Labhasetwar, "Iron oxide nanoparticles for sustained delivery of anticancer agents," *Mol. Pharm.,* vol. 2, no. 3, pp. 194-205, 2005.

[72] F. Scherer F, M. Anton, U. Schillinger, J. Henke, C. Bergemann, A. Kruger, B. Gänsbacher and C. Plank, "Magnetofection: enhancing and targeting gene delivery by magnetic force in vitro and in vivo," *Gene Ther.,* vol. 9, no. 2, pp. 102-109, 2002.

[73] T. K. Jain, J. Richey, M. Strand, D. L. Leslie-Pelecky, C. A. Flask and V. Labhasetwar, "Magnetic nanoparticles with dual functional properties: Drug delivery and magnetic resonance imaging," *Biomaterials,* vol. 29, no. 29, pp. 4012-4021, 2008.

[74] E. Okon, D. Pouliquen, P. Okon, Z. V. Kovaleva, T. P. Stepanova, S. G. Lavit, B. N. Kudryavtsev and P. Jallet, "Biodegradation of magnetite dextran nanoparticles in the rat. A histologic and biophysical study," *Lab. Invest.,* vol. 71, no. 6, pp. 895-903, 1994.

[75] M. Johannsen, U. Gneveckow, L. Eckelt, A. Feussner, N. Waldöfner, R. Scholz, S. Deger, P. Wust, S. A. Loening and A. Jordan, "Clinical hyperthermia of prostate cancer using magnetic nanoparticles: presentation of a new interstitial technique," *Int. J.Hyperthermia,* vol. 21, no. 7, pp. 637-647, 2005.

[76] A. S. Lubbe, C. Bergemann, H. Riess, F. Schriever, P. Reichardt, K. Possinger, M. Matthias, B. Dorken, F. Herrmann, R. Gurtle, P. Hohenberger, N. Haas, R. Sohr, B. Sander, A. J. Lemke, D. Ohlendorf, W. Huhnt and D. Huhn, "Clinical experiencess with magnetic with targheting a phase I study with 4 epidoroxibicin with 14 patients with advanced solid tumors," *Cancer Res.,* vol. 56, no. 20, p. 4686–4693, 1996.

[77] Y. Zhang, J. Y. Liu, F. Yang, Y. J. Zhang, Q. Yao and T. Y. Cui, "A new strategy for assembling multifunctional nanocomposites with iron oxide and amino-terminated PAMAM dendrimers," *J. Mater Sci: Mater Med,* vol. 20, no. 12, pp. 2433-2440, 2009.

[78] Y. M. Huh, Y. W. Jun, H. T. Song, S. Kim, J. S. Choi, J. H. Lee, S. Yoon, K. S. Kim, J. S. Shin, J. S. Suh and J. Cheon, "In vivo magnetic resonance detection of cancer by using multifunctional magnetic nanocrystals," *J. Am. Chem. Soc.,* vol. 127, no. 35, pp. 12387-12391, 2005.

[79] Z. Medarova, W. Pham, Y. Kim Y, G. P. Dai and A. Moore, "In vivo imaging of tumor response to therapy using a dual-modality imaging strategy," *Int. J.Cancer,* vol. 118, no. 11, pp. 2796-802, 2006.

[80] J. Dobson, "Nanoscale biogenic iron oxides and neurodegenerative disease," *FEBS Letters,* vol. 496, no. 1, pp. 1-5, 2001.

[81] Q. Pankhurst, D. Hautot, N. Khan and J. Dobson, "Increased Levels of Magnetic Iron Compounds in Alzheimer's Disease," *Journal of Alzheimer"s Disease,* vol. 13, no. 1, pp. 49-52, 2008.

[82] R. Y. Kwong, Cardiovascular Magnetic Resonance Imaging, Totowa, New Jersey: Humana Press, 2008.

[83] M. Lewin, N. Carlesso, C. H. Tung, X. W. Tang, D. Cory, D. T. Scadden and R. Weissleder, "Tat peptide-derivatized magnetic nanoparticles allow in vivo tracking and recovery of progenitor cells," *Nat Biotechnol.,* vol. 18, no. 4, pp. 410-4, 2000.

[84] D. Sellmyer and R. Skomski, "Introduction" to Advanced Magnetic Nanostructures, Nebraska Lincoln: *Springer,* 2006.

[85] C. Alexiou, W. Arnold, P. Huh, R. J. Klein, H. Renz, F. Parak, C. Bergemann and A. Lubbe, "Magnetic mitoxantrone nanoparticle detection by histology, X-ray and MRI after magnetic tumor targeting," *J. Magn. Magn. Mater.,* vol. 225, pp. 187-193, 2001.

[86] C. Gruttner, S. Rudershausen and J. Teller, "Improved properties of magnetic particles by combination of different polymer materials as particle matrix," *J. Magn. Magn. Mater.,* vol. 225, no. 1-2, pp. 1-7, 2001.

[87] C. Gruttner, J. Teller, W. Schutt and F. Westphal, "Scientific and Clinical Applications of Magnetic Carriers," *Eds U. Hafeli* et al. *Plenum Press, New York,* p. 53, 1997.

[88] G. A. van Ewijk, G. J. Vroege and A. P. Philipse, "Convenient preparation methods for magnetic colloids," *J. Magn. Magn. Mater. 201, 31,* vol. 201, pp. 31-33, 1999.

[89] C. Gruttner and J. Teller, "New types of silica-fortified magnetic nanoparticles as tools for molecular biology applications," *J. Magn. Magn. Mater.,* vol. 194, pp. 8-15, 1999.

[90] J. Connolly, T. G. S. Pierre, M. Rutnakornpituk and J. S. Riffle, "Silica coating of cobalt nanoparticles increases their magnetic and chemical

stability for biomedical applications," *Eur. Cells. Mater.,* vol. 3, no. 2, pp. 106-109, 2002.

[91] S. Mornet, F. Grasset, J. Portier and E. Duguet, "Magnetite@silica nanoparticles for biological applications," *Eur. Cells. Mater.,* vol. 3, no. 2, pp. 110-113, 2002.

[92] H. H. Boennemann, W. J. Brijoux, R. Brinkmann, N. Matoussevitch and N. Waldoefner, *Abs. Pap. Am. Chem. Soc. 226, U349,* 2003.

[93] D. L. Thorek, A. K. Chen, J. Czupryna and A. Tsourkas, "Superparamagnetic Iron Oxide Nanoparticle Probes for Molecular Imaging," *Annals of Biomedical Enginneering,* vol. 34, no. 1, pp. 23-38, 2006.

[94] Y. Wang, S. Hussain and G. Krestin, "Superparamagnetic iron oxide contrast agent: physiochemical characteristics and applications in MR imaging," *Eur. Radiol.,* vol. 11, no. 11, pp. 2319-2331, 2001.

[95] R. Weissleder, G. Elizondo, J. Wittenberg, C. A. Rabito, H. H. Bengele and L. Josephson, *Radiology 175,* p. 489–493, 1990.

[96] R. Edelman, "Contrast-enhanced MR imaging of the heart: overview of the literature," *R. Radiology,* vol. 232, p. 653–668, 2004.

[97] S. Mornet, J. Portier and E. Duguet, "A method for synthesis and functionalization of ultrasmall superparamagnetic covalent carriers based on maghemite and dextran," *Journal of Magnetism and Magnetic Materials 293,* p. 127–134, 2005.

[98] J. Bulte, R. Brooks and B. Moskowitz, "Relaxometry and magnetometry of the MR contrast agent MION-46L.," *Magn.Reson. Med. 42,* vol. 42, no. 2, pp. 379-84, 1999.

[99] . Anzai, M. Prince, T. Chenevert, J. H. Maki, F. Londy, M. London and S. J. McLachlan, "MR angiography with an ultrasmall superparamagnetic iron oxide blood pool agent," *J. Magn.Reson. Imaging,* vol. 7, no. 1, p. 209, 1997.

[100] R. Weissleder, J. Heautot, B. Schaffer, N. Nossiff, M. I. Papisov, A. J. Bogdanov and T. J. Brady, "MR lymphography: study of a high-efficiency lymphotrophic agent.," *Radiology,* vol. 191, no. 1, pp. 225-230, 1994.

[101] R. Weissleder, A. Lee, A. Fischman, P. Reimer, T. Shen, R. Wilkinson, R. J. Callahan and T. J. Brady, "Polyclonal human immunoglobulin G labeled with polymeric iron oxide: antibody MR imaging," *Radiology,* vol. 181, no. 1, pp. 245-9, 1991.

[102] L. Remsen, C. McCormick, S. Roman-Goldstein, G. Nilaver, R. Weissleder, Bogdanov A, I. Hellström, R. A. Kroll and E. A. Neuwelt,

"MR of carcinoma-specific monoclonal antibody conjugated to monocrystalline iron oxide nanoparticles: the potential for noninvasive diagnosis," *Am. J. Neuroradiol.,* vol. 17, no. 3, pp. 411-8, 1996.

[103] M. Kresse, S. Wagner, D. Pfefferer, R. Lawaczeck, V. Elste and W. Semmler, "Targeting of ultrasmall superparamagnetic iron oxide (USPIO) particles to tumor cells in vivo by using transferrin receptor pathways," *Magn. Reson. Med.,* vol. 40, no. 2, pp. 236-42, 1998.

[104] C. Jung, "Surface properties of superparamagnetic iron oxide MR contrast agents: ferumoxides, ferumoxtran, ferumoxsil," *Magn. Reson. Imaging,* vol. 13, no. 5, p. 675, 1995.

[105] E. Groman and, L. Josephson, "Low molecular weight carbohydrates as additives to stabilize metal oxide compositions". US Patent 5,248,492, 1993.

[106] K. G. Paul, T. B. Frigo, J. Y. Groman and E. V. Groman, "ynthesis of ultrasmall superparamagnetic iron oxides using reduced polysaccharides," *Bioconjugate Chemistry,* vol. 15, no. 2, pp. 394-401, 2004.

[107] E. Duguet, S. Mornet and J. Portier, "Ferrofluids stable in neutral media and ferrofluids employing surface-modified particles". French Patent FR 2,855,315 Patent PCT WO2004107368, 2004.

[108] M. R. Carroll, R. C. Woodward, M. J. House, W. Y. Teoh, R. Amal, T. L. Hanley and T. G. St Pierre, "Experimental validation of proton transverse relaxivity models for superparamagnetic nanoparticle MRI contrast agents," *Nanotechnology,* vol. 21, no. 3, p. 035103, 2010.

[109] X. Xuan and C. Zhang, "Controllable Assembly of Hydrophobic Superparamagnetic Iron Oxide Nanoparticle withm PEG-PLA Copolymer and Its Effect on MR Transverse Relaxation Rate," *Journal of Nanomaterials,* 2011.

[110] C. Bean and J. Livingston, "Superparamagnetism,," *Journal of Applied Physics, 30,* pp. S120-S129, 1959.

[111] J. Berret, N. Schonbeck and F. Gazeau, "Controlled clustering of superparamagnetic nanoparticles using block copolymers: design of new contrast agents for magnetic resonance imaging,," *Journal of the American Chemical Society, vol. 128,* p. 1755–1761, 2006.

[112] S. Conroy, S. H. Jerry and M. Z. Lee, "Magnetic nanoparticles in MR imaging and drug delivery," *Advanced Drug Delivery Reviews,* vol. 60, pp. 1252-1265, 2008.

[113] T. F. Massoud and S. S. Gambhir, "Molecular imaging in living subjects: seeing fundamental biological processes in a new light," *Genes Dev,* vol. 17, no. 5, pp. 545-80, 2003.

[114] R. Weissleder, "Molecular Imaging in Cancer," *Science,* vol. 312, pp. 1168-1172, 2006.

[115] J. W. M. Bulte and D. L. Kraitchman, "Iron oxide MR contrast agents for molecular and cellular imaging," *NMR in Biomedicine,* vol. 17, no. 7, p. 484–499, 2004.

[116] H. Gupta and R. Weissleder, "Targeted contrast agents in MR imaging," *Magn Reson Imaging,* vol. 4, no. 1, pp. 171-184, 1996.

[117] K. P. Li and M. D. Bednarski, "Vascular-targeted molecular imaging using functionalized polymerized vesicles," *Journal of Magnetic Resonance Imaging,* vol. 16, no. 4, p. 388–393, 2002.

[118] J. W. M. Bulte, S. C. Zhang, P. van Gelderen, V. Herynek, E. K. Jordan, I. D. Duncan and J. A. Frank, "Neurotransplatantion of magnetically labeled oligodendrocyte progentitors: magnetic resonance tracking of cell migration and myelination," *Proc. Natl. Acad. Sci.,* vol. 96, pp. 15256-15261, 1999.

[119] M. Hoehn, E. Kustermann, J. Blunk, D. Wiedermann, T. Trapp, S. Wecker, S. Focking, H. Arnold, J. Hescheler, B. K. Fleischmann, W. Schwindt and C. Buhrle, "Monimitoring of implanted stem cell migration in vivo: a higly resolved in vivo magnetic resonance imaging investigation of experimental stroke in rat," *Proc. Natl. Acad. Sci. U.S.A.,* vol. 99, pp. 16267-1672, 2002.

[120] E. Shapiro, S. Stanko, K. Sharer, M. Jonathan, C. Dunbar and A. Koretsky, "MRI detection of single particles for cellular imaging," *PNAS,* vol. 101, no. 30, p. 10901–10906, 2004.

[121] K. Hinds, J. Hill, E. Shapiro, M. Laukkanen, A. Silva, C. Combs, T. Varney, R. Balaban, A. Koretsky and C. Dunbar, "Highly efficient endosomal labeling of progenitor and stem cells with large magnetic particles allows magnetic resonance imaging of single cells," *Blood,* vol. 102, p. 867–872, 2003.

[122] M. McAteer, N. Sibson, C. von zur Mühlen, J. Schneider, A. Lowe, N. Warrick, K. Channon, D. Anthony and R. Choudhury, "In vivo magnetic resonance imaging of acute brain inflammation using microparticles of iron oxide," *Nat. Med.,* p. 1253–125, 2007.

[123] K. S. Mankia, M. A. McAteer and R. P. Choudhury, "Microparticle-Based Molecular MRI of Atherosclerosis, Thrombosis, and Tissue Ischemia," *Curr Cardiovasc Imaging Rep.,* p. 17–23, 2011.

[124] Y. Wu, Q. Ye and L. Foley, "In situ labelling of immune cells with iron oxide particles: an approach to detect organ rejection by cellular MRI," *Proc. Natl. Acad. Sci. USA,* pp. 1852-1857, 2006.

[125] Q. Ye, Y. Wu and L. e. a. Foley, "Longitudinal tracking of recipient macrophages in a rat chronic cardiac allograft rejection model with non-invasive magnetic resonance imaging using micrometer-sized paramagnetic iron oxide particles," *Circulation,* p. 118:14, 2008.

[126] M. McAteer, J. Schneider and Z. e. a. Ali, "Magnetic resonance imaging of endothelial adhesion molecules in mouse atherosclerosis using dual-targeted microparticles of iron oxide," *Atheroscler Thromb Vasc Biol,* pp. 77-83, 2008.

[127] . C. Wu and I. Y. Chen, "Cardiovascular Molecular Imaging: Focus on Clinical Translation," *Circulation,* vol. 123, no. 4, p. 425–443, 2011.

[128] J. B. Mathieu and S. Martel, "Magnetic steering of iron oxide microparticles using propulsion gradient coils in MRI," in *Proceedings of the 28th IEEE EMBS Annual International Conference*, New York City, USA,, 2006.

[129] M. Chirita, R. Banica, A. Dabici and I. Grozescu, "Synthesis procedure of (alpha-Fe2O3) nanoparticle by hydrothermal decomposition of Fe-EDTA complex". Romania Patent Patent proposal, no.A/00801, 2009.

[130] M. Chirita, M. L. Kiss, A. Ieta, A. Ercuta and I. Grozescu, "Synthesis of Micrometric Single Crystalline Magnetite with Superparamagnetic Properties for Biomedical Applications," in *Nano Science and Technology Institute-Nanotech 2013, www.nsti.org, ISBN 978-1-4822-0581-7 Vol 1,*, Washington DC, U.S.A., 2013.

[131] M. L. Kiss, M. Chirita, C. A. Beljung, R. Polanek, C. Savii, A. Ieta and A. I. Mihaila Chirita, "Single Crystalline Micrometric Iron Oxide Particles with Superparamagnetic Behaviour for MRI Apllications," in *Nano Science and Technology Institute-Nanotech 2014*, Washigton, 2014.

[132] B. N. Hyon, C. S. In and T. Hyeon, "Inorganic Nanoparticles for MRI Contrast Agents," *Adv. Mater.,* vol. 21, p. 2133–2148, 2009.

[133] L. M. Imaging, "bloodpoolagents," Radcliffe Publishing, 2014. [Online]. Available: URL: http://www.bloodpoolagents.us/content/magnetic-resonance-angiography. [Accessed 20 05 2014].

[134] Y. Wang, S. Hussain and G. Krestin, "Superparamagnetic iron oxide contrast agent: physiochemical characteristics and applications in MR imaging.," *Eur Radiol.,* pp. 2319-2331, 2001.

[135] R. H. Müller, M. Lück, S. Harnisch and K. Thode, "Intravenously Injected Particles; Surface Properties and Interaction with Blood Proteins - the Key Determining the Organ Distribution," in *Scientific and Clinical Applications of Magnetic Carriers (Schütt, W., Teller, J., Häfeli, U., Zborowski, M., eds.),*, 1997, pp. 135-148.

[136] J. Kreuter, "Drug targeting with nanoparticles," *Eur. J. Drug Metab. Pharmacokinet,* vol. 19, p. 253, 1994.

[137] H. Thompsen, T. Almen and S. Morcos, "Contrast Media Safety Committee of the European Society of urogenital radiology E. Gadolinium," *Eur. Radiol.,* pp. 2600-2605, 2002.

[138] P. Majewski and B. Thierry, "Functionalized Magnetite Nanoparticles-Synthesis, Properties, and Bio-Applications," *Critical Reviews in Solid State and Materials Sciences,* vol. 32, no. 3-4, pp. 203-215, 2007.

[139] Y. W. Jun, Y. M. Huh, J. S. Choi, J. H. Lee, H. T. Song, S. Kim, S. Yoon, K. S. Kim, J. S. Shin, J. S. Suh and J. Cheon, "Nanoscale size effect of magnetic nanocrystals and their utilization for cancer diagnosis via magnetic resonance imaging," *J. Am. Chem. Soc,* vol. 127, no. 16, pp. 5732-3, 2005.

[140] J. H. Lee, Y. M. Huh, Y. W. Jun, J. W. Seo, J. T. Jang, H. T. Song, S. Kim, E. J. Cho, H. G. Yoon, J. S. Suh and J. Cheon, "Artificially engineered magnetic nanoparticles for ultra-sensitive molecular imaging," *Nat. Med.,* vol. 13, no. 1, pp. 95-9, 2007.

[141] G. F. Goya, T. S. Berquo, F. C. Fonseca and M. P. Morales, "Static and dynamic magnetic properties of spherical magnetite nanoparticles," *J. Appl. Phys. 94,* 2003.

[142] M. G. Harisinghani, J. Barentsz, P. F. Hahn, S. Deserno, W. M. Deserno, S. Tabatabaei, C. Van De Kaa, J. De La Rosette and R. Weissleder, "C. H. Van De Kaa, J. De La Rosette, and R. Weissleder, Noninvasive detection of clinically occult lymph-node metastases in prostate cancer," *J. Med. 348,* 2003.

[143] J. Lee, M. A. Smith, W. Liu, E. MGold, B. Lewis, H. Song, A. Frank, „Enhanced stem cell tracking via electrostatically assembled fluorescent SPION–peptide complexes," *Nanotechnology 20,* 2009.

[144] A. S. Arbab, G. T. Yocum, H. Kalish, E. K. Jordan, S. A. Anderson, A. Y. Khakoo, E. Read, J. Frank, *Blood 104,* pp. 1217-, 2004.

[145] J. H. Lee, B. Schneider, E. K. Jordan, W. Liu and J. A. Frank, *Adv. Mater. 20,* p. 2512–16, 2008.

[146] I. J. M. de Vries, W. J. Lesterhuis, J. O. Barentsz, P. Verdijk, J. H. van Krieken, O. C. Boerman, W. J. Oyen, J. J. Bonenkamp, G. J. Adema, J.

W. M. Jeff W M Bulte, T. W. J. Scheenen, C. J. A., Cornelis J A Punt, C. J. A. Punt, A. Heerschap and C. G. Figdor, "Magnetic resonance tracking of dendritic cells in melanoma patients for monitoring of cellular therapy," *Nature Biotechnology,* vol. 23, pp. 1407-1413, 2005.

[147] D. L. Kraitchman, A. W. Heldman, E. Atalar, L. C. Amado, B. J. Martin, M. F. Pittenger, J. M. Hare, J. M. Hare and J. M. W. Bulte, "In Vivo Magnetic Resonance Imaging of Mesenchymal Stem Cells in Myocardial Infarction," *Circulation,* vol. doi: 10.1161/ 01.CIR.0000070931.62772.4E, 2014.

[148] R. Zhou, P. D. Acton and V. A. Ferrari, "Imaging Stem Cells Implanted in Infarcted Myocardium," *J. Am Coll Cardiol.,* vol. 48, no. 10, p. 2094–2106, 2006.

[149] P. Tartaj, M. d. P. Morales, S. Veintemillas-Verdaguer, T. González-Carreno and C. J. Serna, "The preparation of magnetic nanoparticles for applications in biomedicine," *Journal of Physics D: Applied Physics,* vol. 7, no. 7, pp. doi:10.1088/0022-3727/36/13/202, 2003.

[150] A. A. Faridani, E. L. Ritman and K. T. Smith, "Examples of local tomography," *J. Appl. Math,* vol. 52, no. 4, pp. 1193-1198, 1992.

[151] C. Savii and A. Putz, "Recent Advances in Bioresponsive Nanomaterials," in *Carbon Bonding and Structures:Advances in Physics and Chemistry*, London, Springer Verlag, 2011, pp. 379-435.

[152] C. Savii, C. Ionescu, M. Popovici, C. Enache and R. Radu, "Nonionic Organic-Inorganic Hybrid Materials," in *New Trends in Nonionic (Co)Polymers and Hybrids*, New York, Nova Science Publishe, 2006, pp. 215-261.

[153] G. A. Hughes, "Nanostructure-mediated drug delivery," *Nanomedicine: Nanotechnology, Biology, and Medicine,* vol. 1, no. 1, pp. 22-30, 2005.

[154] U. O. Häfeli, G. J. Pauer, W. K. Roberts, J. L. Humm and R. M. Macklis, "Magnetically targeted microspheres for intracavitary and intraspinal Y-90 radiotherapy. In: Häfeli U, Schütt W, Teller J, Zborowski M, editor," in *Scientific and clinical applications of magnetic carriers*, New York, Plenum, 1997, pp. 501-516.

[155] T. Neuberger, B. Schopf, H. Hofmann, M. Hofmann and B. Rechenberg, "Superparamagnetic nanoparticles for biomedical applications: Possibilities and limitations of a new drug delivery system," *J. Magnetism Magnetic Mater.,* 2005.

[156] N. Kohler, C. Sun, J. Wang and M. Zhang, "Methotrexate-modified superparamagnetic nanoparticles and their intracellular uptake into human cancer cells," *Langmuir,* vol. 21, no. 19, p. 8858, 2005.

[157] A. Gupta and M. Gupta, "Synthesis and surface engineering of iron oxide nanoparticles for biomedical applications," *Biomaterials 26,* p. 3995–4021, 2005.

[158] Z. Medarova, W. Pham, C. Farrar, V. Petkova and A. Moore, "In vivo imaging of siRNA delivery and silencing in tumors.," *Nat. Med.,* vol. 13, no. 3, pp. 372-7, 2007.

[159] X. Wang, R. Zhang, C. Wu, Y. Dai, M. Song, S. Gutmann, F. Gao, G. Lv, J. Li, X. Li, Z. Guan, D. Fu și B. Chen, „The application of Fe(3)O(4) nanoparticles in cancer research: A new strategy to inhibit drug resistance," *J. Biomed. Mater. Res. A,* vol. 80, nr. 4, pp. 852-860, 2007.

[160] A. S. Lübbe, C. Bergemann, J. Brock and D. G. McClure, "Physiological aspects in magnetic drug targeting," *J. Magn. Magn. Mater.,* vol. 194, pp. 149-155, 1999.

[161] J. Fricker, "Drugs with a magnetic attraction to tumours," *Drug Discovery Today,* vol. 6, no. 8, p. 387–389, 2001.

[162] A. Lübbe, C. Bergemann, J. Brock and D. McClure, "Physiological aspects in magnetic drug targeting," *J. Magn. Magn. Mater.,* vol. 194, p. 149–155, 1999.

[163] P. Majewski and B. Thierry, "Functionalized Magnetite Nanoparticles-Synthesis, Properties, and Bio-Applications," *Critical Reviews in Solid State and Materials Sciences,* pp. 203 - 215, 2007.

[164] J. M. Gallo, P. K. Gupta, C. Hung and D. Perrier, "Evaluation of drug delivery following the administration of magnetic albumin microspheres containing adriamycin to the rat," *J. Pharm. Sci.,* vol. 78, no. 3, p. 190–194, 1989.

[165] S. K. Pulfer and J. M. Gallo, Scientific and clinical applications of magnetic carriers, New York: Plenum press, 1997.

[166] A. S. Lübbe, C. Alexiou and C. Bergemann, "Clinical Applications of Magnetic Drug Targeting," *J. Surg. Res.,* vol. 95, pp. 200-206, 2001.

[167] A. Mountain, "Gene therapy: the first decade," *Trends Biotechnol,* vol. 18, no. 3, pp. 119-128, 2000.

[168] W. Wang, W. Li, N. Ma and G. Steinhoff, "Non-viral gene delivery methods," *Curr. Pharm. Biotechnol.,* vol. 14, no. 1, pp. 46-60, 2013.

[169] S. Jin, J. C. Leach and K. Ye, "Nanoparticle-mediated gene delivery," *Methods Mol Biol.,* vol. 544, no. doi: 10.1007/978-1-59745-483-4_34, pp. 547-57, 2009.

[170] H. Tian, J. Chen and X. Chen, "Nanoparticles for Gene Delivery," *Small,* vol. 9, no. 12, p. 2034–2044, 2013.

[171] S. Jin, J. C. Leach and K. Ye, "Nanoparticle-mediated gene delivery," *Methods Mol. Biol.,* vol. 544, pp. 547-557, 2009.

[172] S. Purushotham, P. E. J. Chang, H. Rumpel, I. H. C. Kee, R. T. H. Ng, P. K. H. Chow, C. K. Tan and R. V. Ramanujan, "Thermoresponsive core–shell magnetic nanoparticles for combined modalities of cancer therapy," *Nanotechnology,* vol. 20, p. 3051, 2009.

[173] S. B. Field and J. W. Hand, "An Introduction to the Practical Aspectsl of Clinical Hyperthermia," London, Taylor and Francis, 1990, pp. 256-259.

[174] K. Overgaard and J. Overgaard, "Investigations on the possibility of a thermic tumour therapy. I. Short-wave treatment of a transplanted isologous mouse mammary carcinoma," *Eur. J. Cancer,* p. 65–78, 1972.

[175] J. Overgaard, "Effect of hyperthermia on malignant cells in vivo. A review and a hypothesis," *Cancer,* vol. 39, no. 1, p. 2637–2645, 1977.

[176] M. Lévy, C. Wilhelm, J.-M. Siaugue, O. Horner, J. Bacri şi F. Gazeau, „Magnetically induced hyperthermia: size-dependent heating power of γ-Fe(2)O(3) nanoparticles," *J. Phys. Condens Matter,* vol. 20, nr. 20, p. 204133, 2008.

[177] A. Jordan, "First clinical trial of magnetic fluid hyperthermia (MFH) for local tumor treatment," *Scientific and Clinical Applications of Magnetic Carriers,* 2004.

[178] C. Wilhelm and F. Gazeau, "Magnetic nanoparticles: Internal probes and heaters within living cells," *Journal of Magnetism and Magnetic Materials 321,* p. 671–674, 2009.

[179] A. Jordan, P. Wust, H. Fahling, W. John, A. Hinz and R. Felix, "Inductive heating of ferrimagnetic particles and magnetic fluids: physical evaluation of their potential for hyperthermia," *Int. J. Hyperthermia,* vol. 9, no. 1, pp. 51-68, 1993.

[180] R. Rosensweig, "Heating magnetic fluid with alternating magnetic field," *J. Magn. Magn. Mater.,* vol. 252, no. 11, p. 370–374, 2002.

[181] J. Fortin, C. Wilhelm, J. Servais, C. Ménager, J. C. Bacri and F. Gazeau, "Size-sorted anionic iron oxide nanomagnets as colloidal mediators for magnetic hyperthermia," *J. Am. Chem. Soc.,* vol. 129, no. 9, pp. 2628-35., 2007.

[182] C. Lévy, C. Wilhelm, J. M. Siaugue, O. Horner, J. C. Bacri and F. Gazeau, "Magnetically induced hyperthermia: size-dependent heating power of γ-Fe2O3 nanoparticles," *J. Phys.: Condens. Matter,* vol. 20, no. 20, p. 204133, 2008.

[183] R. K. Gilchrist, R. Medal, W. D. Shorey, R. C. Hanselman, J. C. Parrott and C. B. Taylor, "Selective inductive heating of lymph nodes," *Ann. Surg.,* vol. 146, p. 596–606, 1957.

[184] A. S. Lübbe, C. Bergemann, C. W. Huhnt, T. Fricke, H. Riess, J. W. Brock and D. Huhn, "Preclinical experiences with magnetic drug targeting: tolerance and efficacy," *Cancer Res.,* vol. 56, no. 20, pp. 4694-701, 1996.

[185] J. P. Fortin, F. Gazeau and C. Wilhelm, "Intracellular heating of living cells through Néel relaxation of magnetic nanoparticles," *Eur. Biophys. J.,* vol. 37, no. 2, pp. 223-8, 2008.

[186] C. Wilhelm, J. Fortin and F. Gazeau, "Tumour cell toxicity of intracellular hyperthermia mediated by magnetic nanoparticles," *J. Nanosci. Nanotechnol.,* vol. 7, no. 8, pp. 2933-2937, 2007.

[187] A. Ito, M. Fujioka, T. Yoshida, K. Wakamatsu, S. Ito, T. Yamashita, K. Jimbow and H. Honda, "4-S-Cysteaminylphenol-loaded magnetite cationic liposomes for combination therapy of hyperthermia with chemotherapy against malignant melanoma," *Cancer Sci.,* vol. 98, no. 3, p. 42, 2007.

[188] A. Jordan, R. Scholz, P. Wust, H. Schirra, T. Schieste și H. Schmidt, „Endocytosis of Dextran or Silan Coated Superparamagnetic: Magnetite Nanoparticles and the Effect of Intracellular Hyperthermia on Human Mammary Carcinoma Cells in-vitro.," *J. Magn. Magn. Mater.,* vol. 94, pp. 185-196, 1999.

[189] F. Sonvico, S. Mornet, S. Vasseur, C. Dubernet, D. Jaillard, J. Degrouard, J. Hoebeke, E. Duguet, P. Colombo and P. Couvreur, "Folate-conjugated iron oxide nanoparticles for solid tumor targeting as potential specific magnetic hyperthermia mediators: Synthesis, physicochemical characterization, and in vitro experiments," *Bioconjug. Chem.,* vol. 16, no. 5, pp. 1181-8, 2005.

[190] A. Ito, Y. Kuga, H. Honda, H. Kikkawa, A. Horiuchi, Y. Watanabe and T. Kobay, "Magnetite nanoparticle-loaded Anti-Her2 immunoliposomes for combination of antibody therapy with hyperthermia.," *Cancer Lett., 212,* 2004.

[191] P. Wust, U. Gneveckow, M. Johannsen, D. Bohmer, T. Henkel, F. Kahmann, J. Sehouli, R. Felix, J. Ricke and A. Jordan, "Magnetic nanoparticles for interstitial thermotherapy—feasibility, tolerance and achieved temperatures," *Int. J. Hyperthermia,* vol. 22, no. 8, p. 673, 2006.

[192] R. Ivkov, S. J. Denardo, W. Daum, A. R. Foreman, R. C. Goldstein,,V. S. Nemkov, G. L, V. S. Nemkov şi G. Dernardo, „Application of high amplitude alternating magnetic fields for heat induction of nanoparticies localized in cancer," *Clin. Cancer Res. 11,* 2006.

[193] A. P. Khandhar, R. M. Ferguson and K. M. Krishnan, "Monodispersed magnetite nanoparticles optimized for magnetic fluid hyperthermia: Implications in biological systems," *J. Applied Physics,* vol. 109, no. 07B310, 2011.

[194] H. S. Aliabadi, M. Mozaffari, B. Behdadfar, M. Raesizadeh and H. Zarkesh-Esfahani, "Preparation and Cytotoxic Evaluation of Magnetite (Fe3O4) Nanoparticles on Breast Cancer Cells and its Combinatory Effects with Doxorubicin used in Hyperthermia Avicenna," *J. Med. Biotech.,* vol. 5, no. 2, pp. 96-103, 2013.

[195] B. Chen, Q. Sun, X. Wang, F. Gao, Y. Dai, Y. Yin, J. Ding, C. Gao, J. Cheng and J. Li, "Reversal in multidrug resistance by magnetic nanoparticle of Fe3O4 loaded with adriamycin and tetrandrine in K562/A02 leukemic cells," *Int. J. Nanaomedicine,* vol. 3, pp. 277-286, 2008.

[196] O. D. Jayakumar, R. Ganguly, A. K. Tyagi, D. K. Chandrasekharan, N. Krishnan and K. Cherupally, "Water dispersible Fe3O4 nanoparticles carrying doxorubicin for cancer therapy," *J. Nanosci. Nanotechnol.,* vol. 9, pp. 6344-6348, 2009.

[197] M. H. Yazdi, Z. N. Najafi, M. R. Khorramizadeh, M. Amini and A. R. Shahverdi, "Fabrication of docetaxel surfaced Fe3O4 magnetite nanoparticles and their cytotoxicity on 4T1 breast cancer cells," *DARU J. Pharm. Sci.,* vol. 20, pp. 15-19, 2012.

[198] E. Cheraghipour, S. Javadpour and A. R. Mehdizadeh, "Citrate capped superparamagnetic iron oxide nanoparticles used for hyperthermia therapy," *J. Biomedical Science and Engineering,* vol. 5, pp. 715-719, 2012.

[199] E. Neagu, E. Teodor, G. L. Radu, A. C. Nechifor and G. Nechifor, "Immobilization of invertase on magnetic nanoparticles," *Romanian Biological Sciences,* vol. 3, no. 3-4, pp. 1-9, 2005.

[200] A. K. Johnson, A. M. Zawadzka, L. A. Deobald, R. L. Crawford and A. J. Paszczynski, *J. Nanopart. Res.,* vol. 10, pp. 1009-1025, 2008.

[201] D. Bahadur and J. Giri, "Biomaterials and magnetism," *Sadhara,* vol. 28, no. 3-4, p. 639–656, 2003.

[202] D. Horák, M. Babič, P. Jendelová, V. Herynek, M. Trchová, K. Likavčanová, M. Kapcalová, M. Hájek and E. Sykova, "Effect of

different magnetic nanoparticle coatings on the efficiency of stem cell labeling," *Journal of Magnetism and Magnetic Materials 321,* p. 1539–1547, 2009.

[203] S. Zhang, Z. Bian, C. Gu, Y. Zhang, S. He, N. Gu and J. Zhang, "Preparation of anti-human cardiac troponin I immunomagnetic nanoparticles and biological activity assays," *Colloids and Surfaces B: Biointerfaces,* p. 143–148, 2007.

[204] M. I. Shukoor,, F. Natalio, M. N. Tahir, M. Divekar, N. Metz, H. A. Therese, P. Theato, V. Ksenofontov, H. C. Schröder, W. G. Müller and W. Tremel, "Multifunctional polymer-derivatized g- Fe2O3 nanocrystals as a methodology for the biomagnetic separation of recombinant His-tagged proteins," *Journal of Magnetism and Magnetic Materials,* vol. 320, p. 2339– 2344, 2008.

[205] P. L. Kronick, G. L. Campbell and K. Joseph, "Magnetic microspheres prepared by redox polymerization used in a cell separation based on gangliosides," *Science,* vol. 200, no. 4344, pp. 1074-1076, 1978.

[206] I. Safarik, M. Safarikova and S. Forsythe, "The application of magnetic separations in applied microbiology," *J. Appl. Bacteriol.,* vol. 78, no. 6, p. 575–585, 1995.

[207] B. Naume, U. Nonstad, B. Stekinjer, S. Funderud, E. Smeland and T. Espevik, "Immunomagnetic isolation of NK and LAK cells," *J. Immunol. Methods,* vol. 136, no. 1, pp. 1-9, 1991.

[208] T. Osaka, T. Matsunaga, T. Nakanishi, A. Arakaki, D. Niwa and H. Iida, "Synthesis of magnetic nanoparticles and their application to bioassays," *Anal. Bioanal. Chem.,* vol. 384, no. 3, pp. 593-600, 2006.

[209] Y. R. Chemla, H. L. Grossman, Y. Poon, R. McDermott, R. Stevens, M. D. Alper and J. Clarke, "Ultrasensitive magnetic biosensor for homogeneous immunoassay," *Proc. Natl. Acad. Sci. USA,* vol. 97, no. 26, p. 14568, 2000.

[210] J. M. Perez, L. Josephson, O. T. Loughlin, D. Hogemann and R. Weissleder, "Magnetic relaxation switches capable of sensing molecular interactions," *Nat. Biotechnol.,* vol. 20, no. 8, p. 816, 2002.

[211] M. V. Kutushov, A. A. Kuznetsov, V. I. Filippov and O. A. Kuznetsov, in *Scientific and clinical applications of magnetic carriers*, NewYork, London: Plenum, 1997, p. 391.

[212] P. C. Lin, P. H. Chou, S. H. Chen, H. K. Liao, K. Wang, Y. J. Chen and C. C. Lin, "Ethylene glycol-protected magnetic nanoparticles for a multiplexed immunoassay in human plasma," *Small 2,* 2006.

[213] enrichment for detection of carcinoma cells in fluid specimens," *Cancer 94,* 2002.

[214] X. Hong, Y. Liu, J. Li, W. Guo and Y. Bai, "Detection of magnetic-labeled antibody specific recognition events by combined atomic force and magnetic force microscopy," *Journal of Magnetism and Magnetic Materials,* vol. 321, no. 17, p. 2607–2611, 2009.

[215] D. Horák, M. Babič, P. Jendelová, V. Herynek, M. Trchová, K. Likavčanová, M. Kapcalov, M. Sykova, M. Eva Sykova and M. Hájek, "Effect of different magnetic nanoparticle coatings on the efficiency of stem cell labeling," *Journal of Magnetism and Magnetic Materials,* vol. 321, pp. 1539-1547, 2009.

[216] S. Utech, C. Scherer and M. Maskos, "Multifunctional, multicompartment polyorganosiloxane magnetic nanoparticles for biomedical applications," in *Proceedings of the Seventh International Conference on the Scientific and Clinical Applications of Magnetic Carriers*, Vancouver, 2009.

Reviewed by:

Dr. Adrian Ieta, Department of Physics, State University of New York at Oswego, Oswego, NY, Email: adrian.ieta@gmail.com.

In: Magnetite
Editor: Jacqueline Ward

ISBN: 978-1-63321-591-7

Chapter 5

INTERPOLYELECTROLYTE FUNCTIONALIZED SHELLS OF MAGNETITE NANO- AND MICROPARTICLES

N. A. Samoilova*, V.E. Tikhonov, M. A. Krayukhina and I. A. Yamskov

A.N. Nesmeyanov Institute of Organoelement Compounds,
Russian Academy of Sciences, Moscow, Russia

ABSTRACT

A facile method of preparation of stabilized and functionalized nano- and microparticles of magnetite by successive application of oppositely charged polymers with a regular structure of macromolecular chains (chitosan and maleic acid copolymers) onto Fe_3O_4 core are developed. This approach makes it possible to create two types of magnetite interpolyelectrolyte shells, containing carboxylic or amino groups in outer layer of shell. Composition and magnetic properties of composite particles depend on nature of the copolymer of maleic acid, reaction conditions and size of obtained particles. The carboxylic groups of copolymer in outer layer of interpolyelectrolyte shell were converted into

* Corresponding author: Tel./fax.: 495-135-50-37; E-mail: samoilova.nadezhda@ gmail.com.

reactive anhydride groups by heating. Thermal treatment also leads to covalent cross-linking of shell and improves stability of composites both in acidic and alkaline media. Horseradish peroxidase was successfully immobilized onto covalently cross-linked and activated microparticles of Fe_3O_4 in aqueous medium without of condensing agents. The proposed reproducible and low-cost technique does not use toxic reagents or solvents at all stages, including preparation of Fe_3O_4, formation, activation and cross-linking, of magnetite shell, modification of activated surface of composite particles.

1. INTRODUCTION

Magnetic nano- (MNPs) and microparticles (MMPs) especially based on magnetite are widely used in scientific researches, various fields of technology, medicine and biotechnology due to their size effects, availability of precursors, biocompatibility [1-12]. Moreover, magnetic nanomaterials display unusual properties, such as high values of the exchange interaction and abnormally large magnetocaloric effect [13]. Multifunctional properties of these particles allow to use them for various purposes: construction of printing devices in the form of magnetic fluids (ferrofluides) [14], for magnetic storage of information [15], optoelectronic [16], spintronic [17], catalysis [18], removal of hazardous contaminants from aqueous media [19] (for example, copper ions [20], radionuclides [21]), and for medical and biotechnological applications. The most common use of MNPs and MMPs in medicine includes the preparation of magnetic materials for contrast images in magnetic resonance imaging (MRI) [8, 22], systems of directional transportation and delivering of biologically active compounds and drugs [1,4,6,10], construction of three-dimensional “in vivo like” biotissues [23]etc. Biotechnological applications mainly include biosensors preparation [8, 11, 24], immobilization or isolation of drugs, proteins [9, 25-27], nucleic acids [28], cells [8, 29] (including stem cells [30]) etc.

Although sources of magnetite MNPs and MMPs are very diverse (for example, the biogenic magnetite of *Magnetotactic spirillum* is described [31]), usually physical, physico-chemical and chemical methods [5, 7, 9, 12] of MNPs and MMPs preparation have been used mainly. Among them most used ones are thermolysis of precursors and synthesis in solution (microemulsionic method or coprecipitation of Fe (+2) and Fe (+3) salts in an alkaline medium).

After the preparation of magnetite nano- and microparticles their surface requires protection from spontaneous aggregation and corrosion. Polymer coatings allow the modulation of properties of shell – hydrophobization or hydrophilization, functionalization of the surface of particles for desired applications. The developing strategy for construction of stable magnetite shells is the preparation of interpolyelectrolyte complex (IPEC) covering. IPECs, in particular of chitosan (the artificial polyaminosaccharide derived from chitin, available from abundant renewable resources) and anionic polyelectrolytes, are widely used in biotechnology, medicine, pharmacy, food industry, etc.[32] IPECs of chitosan with polymer of monocarboxylic (acrylic) acid as a stabilizing coatings of magnetite have been described [33, 34]. But IPECs of chitosan with alternant copolymers of dicarboxylic (maleic) acid have not been tested for stabilization and functionalization of MNPs and MMPs. The main advantages of maleic acid copolymers are: commercial availability of polymers in form of maleic anhydride copolymers or, at least, possibility of copolymer synthesis according to known procedures; easy to converse maleic anhydride groups to maleic acid ones in aqueous media. Also reverse forming of anhydride groups is possible from maleic acid residues by heating of dried copolymer. Active anhydride groups are capable of reacting with amino and hydroxyl groups of different compounds and participating in a further cross-linking of interpolyelectrolyte shell.

The aim of this study was to develop an efficient technology for the preparation of functionalized, cross-linked and activated magnetite MNPs and MMPs, applicable for biotechnological usages, by means of formation and transformation of interpolyelectrolyte shell of magnetite. To realize the aim the successive application of oppositely charged polymers (chitosan and maleic acid copolymers) onto magnetite surface was used. Two types of coatings were proposed: copolymer of maleic acid was exposed on the outer surface of coating (Figure 1, *Type 1*) and chitosan was exposed on the outer surface of shell (Figure 1, *Type 2*). It was shown that the reinforcement and activation of magnetite shell was achieved by thermal treatment. Composition, pH-stability and magnetic properties of MNPs and MMPs were investigated as well. Also, derivatization protocol was elaborated and possible biotechnological application of MMPs was discussed.

Functionalized nano- and microparticles of magnetite were characterized by FT-IR, TEM, optical microscopy, AFM, DLS, UV-VIS spectrometry, and magnetometry.

2. Experimental

Materials and Characterization Methods

Poly (ethylene-*alt*-maleic anhydride) M 25000 was purchased from Monsanto (USA). Poly (N-vinyl-2-pyrrolidone-*alt*-maleic anhydride), M 40000, was prepared following the described procedure [35]. The copolymers were hydrolyzed to corresponding copolymers of maleic acid by dissolving in water followed by lyophilization. Chitosan (deacetylation degree 85%, M 90000) was purchased from "Bioprogress" (Moscow, Russia). N^{α}–Fluorenilmethyloxycarbonyl-L-Lys hydrochloride and sulforodamin B sodium salt were purchased from Fluka. Methylene blue and Eosin B were the products of Aldrich. Horseradish peroxidase (HRP) was purchased from Sigma (250 U/mg).

Composition of the magnetite composites was determined by elemental Fe, C, H, N analysis (the Laboratory for microanalysis of INEOS RAS). FT-IR spectra of MNPs and MMPs were recorded with a Nicolet Magna IR-720 Fourier transform IR spectrometer (USA). Vibrating sample magnetometer (BHV-55) was used to characterize the magnetic saturations of MNPs and MMPs. UV-VIS absorption spectra were obtained with a UVIKON-922 (BRD) spectrophotometer. Transmission electron microscopy (TEM) micrographs were performed with a LEO 912 AB (Omega, Karl Zeiss; BRD) microscope operating at the accelerating voltage of 100 kV. For TEM observations, a drop of particles colloidal dispersion in water was placed onto Formvar coated copper grid and then dried. Particle size distributions were obtained by measuring of 100 – 300 individual particles. AFM topographic images of MNPs were obtained in tapping mode with a FemtoScan atomic force microscope (Advanced Technologies Center, Russia) with the use of commercially available Si cantilevers (NSG 11 S) with a spring constant of 11.5 N/m. The images were taken at typical scan rate of 0.5 – 1.0 Hz and processed with FemtoScan online software. DLS experiments were conducted on an ALV/DLS/SLS-5022F spectrometer (ALV-GmbH, Langen, Germany) equipped with an ALV5000 digital time correlator and 22 mW UNIPHASE He-Ne laser (λ_0 = 632 nm). Measurements were made at 25 °C. Optical microscopy was performed using an Eclise H550S (Nikon, Japan) microscope, equipped with a Kodak DC 120 Digital Camera. pH-Stability of magnetite polymer coatings was studied using a spectrofluorimeter “Fluorat®-02-panorama” (Lumex, S.-Peterburg, Russia).

Synthesis of Fe_3O_4 Nanoparticles

Magnetite (Fe_3O_4) was prepared by dehydration and coprecipitation of freshly prepared Fe^{+2} and Fe^{+3} hydroxides formed from their salts (taken in 1:2 molar ratio) in an alkaline solution following the described method [36]. Briefly: 1.4g $Fe_2(SO_4)_3{\cdot}9H_2O$ and 0.7g $FeSO_4{\cdot}12H_2O$ were dissolved in 25 mL of Deionized (DI) water, then 10 mL 25% aqueous ammonium solution in 40 mL of water was added under sonication (22 KHz) of reaction mixture. Then the reaction mixture was sonificated for 2 min. The black precipitate of magnetite was collected with the use of a permanent magnet (the magnetic field strength 2000G), washed using DI water several times until pH neutrality of the effluent and was used immediately after the preparation. Obtained dispersion (1.80 g) contained approximately 19% of dry Fe_3O_4.

Preparation of Magnetite/Polymer Composites

Preparation of magnetite/chitosan composite (Fe_3O_4/CS): freshly prepared magnetite (0.50 g dispersion in water, containing 0. 11g of dry Fe_3O_4) was mixed with 50 mL DI water, sonicated (22 KHz) for 2 min and then 5 mL of 2% chitosan solution in CH_3COOH (1%) was added to the mixture. The reaction mixture was additionally sonicated for 2 min. Magnetic composite was collected with the use of the permanent magnet and washed using DI water. Nanoparticles of Fe_3O_4/CS composite (MNPs) were separated from the microparticles (MMPs) by centrifugation at 3000 rpm for 5 min and then MNPs and MMPs were lyophilized. The yield of MNP fraction was 6%, the yield of MMPs – 94%.

Magnetite/poly(ethylene-*alt*-maleic acid) (Fe_3O_4/EM) and magnetite/ poly(N-vinyl-2-pyrrolidone-*alt*-maleic acid) (Fe_3O_4/VM) composites were prepared following the above described protocol with the use of 5 mL 2% copolymer solution in water (pH 5).

Preparation of magnetite/chitosan/copolymer composites (Fe_3O_4/CS/EM): Fe_3O_4/CS MNPs or MMPs (0.10 g), were dispersed in 50 mL of DI water, sonicated (22 KHz) for 2 min and then 1 mL 2% aqueous solution of EM (pH 5) was added to the mixture. The reaction mixture was sonicated for 2 min. The composites were collected with the use of the permanent magnet, washed using DI water and lyophilized. Yield of MMPs was 0.11g, MNPs - 0.14g.

MNPs or MMPs of Fe_3O_4/EM/CS or Fe_3O_4/VM/CS were prepared following the above described protocol with the use of Fe_3O_4/EM or

Fe_3O_4/VM and 1 mL 2% chitosan solution in CH_3COOH (1%). Yield of MMPs fractions was 0.12g and 0.13g, respectively.

IPEC CS/EM as the model compound for FT-IR spectra was prepared by the mixing of the aqueous solutions of equimolar amounts of polycation (CS) and polyanion (EM) at pH 5. The precipitate was washed with water and lyophilized.

Evaluation of Content of Outer Polymer Layer in Interpolyelectrolyte Shells by Dye Displacement Method

Content of polymer in outer layer of interpolyelectrolyte shell of nano- and microcomposite particles Fe_3O_4/CS/VM was assessed by dye displacement method. The dye - methylene blue (MB), - from complex VM–MB was displaced by chitosan of Fe_3O_4/CS composite. Complex VM–MB was obtained as follows: aqueous solution of copolymer (0.2 mL, 0.7 x 10^{-3} mol/L) was added to aqueous solution of MB (1.8 mL, 2.0 x 10^{-5} mol/L). Optical density at $\lambda = 665$ nm was measured and was admitted as the initial optical density (D_o). VM–MB complex was mixed with precisely weighted quantities of Fe_3O_4/CS, taken in increasing amounts, and the optical density D_i of the solution containing MB (released from VM–MB) was measured. Following the above described protocol the content of outer chitosan layer in Fe_3O_4/MAC/CS was evaluated by eosin (E) displacement from CS–E complex by precisely weighted quantities of Fe_3O_4/VM or Fe_3O_4/EM, respectively, in aqueous solution at $\lambda = 515$ nm.

For evaluation of the amount of functional groups which are not involved into interpolyelectrolyte interaction the titration of MMPs by the anionic (eosin) or cationic (methylene blue) dyes was carried out. Under titration the increasing volumes of solution of dye were added to precisely weighted amounts of magnetite composite until the process of discoloring of solution had stopped.

Stability Tests of Composites

Stability of MMPs containing labeled polymers in the shell was determined by a fluorimetric method. The parent polymers were labeled and then labeled polymers were applied for magnetite covering.

Preparation of N^{α}–Fluorenilmethyloxycarbonyl(FMOC)-L-Lys -labeled poly(ethylene-*alt*-maleic acid) (FMOC–EM): EM solution (1.72 g, 13.7 mmol) in 15 mL of dry dimethylformamide (DMF) was mixed with the solution of N^{α} –FMOC–L–lysine (0.095 g, 0.137 mmol) in 2 mL of DMF. The resultant solution was stirred for 5 h at room temperature. Then, the solution was diluted with DI water, dialyzed against DI water (5Lx3), and lyophilized. The degree of substitution was calculated from the absorbance of FMOC-EM solution in DI water at $\lambda = 300$ nm and found to be 1 mol%.

Chitosan was labeled by Sulforodamin B (SR) following the described protocol [37] (CS–SR), and the degree of labeling for CS–SR was found to be 0.005 mol. %. Magnetite composite containing labeled polymer (10 mg) was placed in 3 ml buffer (0.05M Glycine–HCl pH 2 or 0.05M Glycine–NaOH pH 9) and the mixture was kept for period 1 – 500 h at 30°C. The MMPs were periodically separated with the use of the permanent magnet, and fluorescent response of the supernatant was measured at $\lambda_{em} = 586$ nm for CS–SR ($\lambda_{ex} = 568$ nm) or at $\lambda_{em} = 260$ nm for FMOC–EM ($\lambda_{ex} = 306$ nm). We analyzed the changes in the fluorescence spectra occurring in the supernatant solution after separation of magnetic-field component from the reaction mixture with the use of the permanent magnet, and then calculated the degree of destruction of MMPs shells. As the control labeled polymers containing MMPs were stored in distilled water (pH 7) at room temperature for period 6 months.

Thermal Activation and Crosslinking of Magnetite Shells, HRP Immobilization onto Activated Composite Particles

For activation and crosslinking of magnetite interpolyelectrolyte shell, a powdery sample of MMPs Fe_3O_4/CS/EM (0.030 g) was heated at 110°C for 3 h over P_2O_5 (under vacuum) and then cooled to room temperature. For FT-IR spectrum the model compound (IPEC CS/EM) was heated under the same conditions.

The activated MMPs were added to 1 mL solution of HRP (2 mg/mL) in Milli-Q quality grade water. The mixture was vigorously stirred for 5 h, and the magnetic particles were collected and washed successively with the water until the absence of peroxidase activity in the effluent, using the permanent magnet for MMPs collection. The contents of immobilized HRP onto activated MMPs was calculated, and peroxidase activity of native and immobilized HRP was determined at $\lambda = 420$ nm by the observing of the oxidation of ferrocyanide $K_4Fe(CN)_6$ to ferricyanide $K_3Fe(CN)_6$ by hydrogen peroxide following the described method [38].

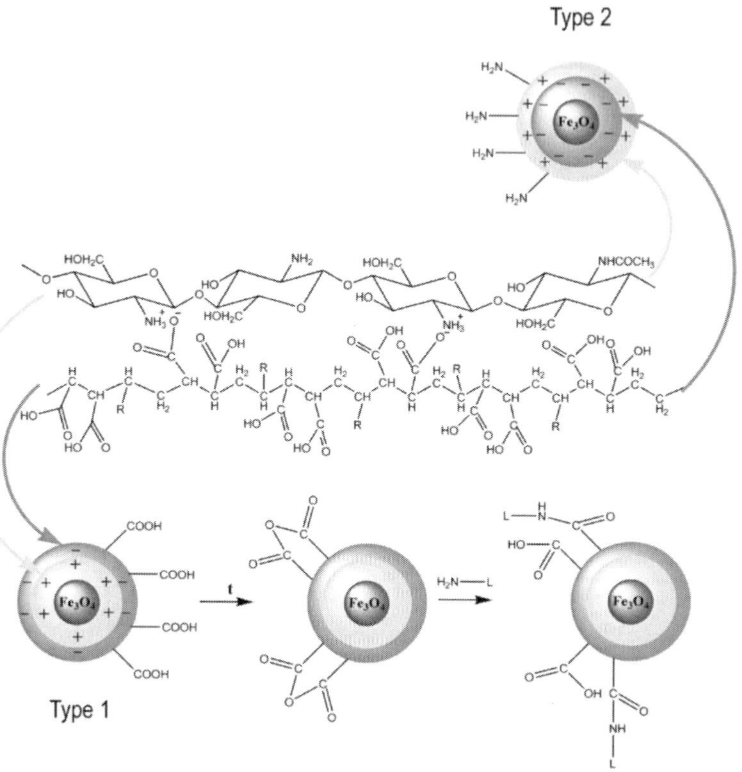

* - Two types of interpolyelectrolyte coatings of magnetite were obtained by successive in two steps application of oppositely charged polymers: the shell of the lower composite (*Type 1*) includes an inner layer of chitosan and outer layer of maleic acid copolymer; an upper composite (*Type 2*) includes the maleic acid copolymer as an inner shell layer and chitosan as an outer one. The interpolyelectrolyte shell of *Type 1* composite was cross-linked thermally and maleic acid residues of copolymer were converted to maleic anhydride ones, then the amine-containing ligand was introduced into the composite (reactions below).

Figure 1. Schematic representation of the method used for magnetite coating, functionalization and modification.*

3. Results and Discussion

Magnetite nanoparticles obtained by the well-known method of coprecipitation of Fe (+2) and Fe (+3) salts in an alkaline medium [36] were covered by polymers immediately after the preparation. In this research the surface stabilization and functionalization of magnetite (Fe_3O_4) particles were achieved by the formation of interpolyelectrolyte shell on Fe_3O_4 core. Surface activation and stronger fixation of the shell were achieved by thermal treatment of composite particles (Figure 1).

Two types of reversed interpolyelectrolyte shells were obtained by the reversed order of application of oppositely charged polymers to magnetite particles. Each type of interpolyelectrolyte coating was formed in two stages: a) formation of inner polymer layer ("monopolymer" covering) and b) creation of interpolyelectrolyte shell.

Formation of Magnetite/Polymer Composites

Nanoparticles of magnetite stabilized with one of hydrophilic polymers (chitosan or poly(acrylic acide)) are well-known. Magnetite nanoclusters stabilized with poly(acrylic acid) have been synthesized in organic phase at 220°C in a nitrogen atmosphere [39]. It has been shown that a part of carboxylic groups of poly(acrylic acid) are strongly coordinated with cations of the magnetite surface, while uncoordinated carboxylic groups extended into the aqueous solution [39]. Chelation of the amino groups of chitosan with polar Fe_3O_4 nanoparticles was confirmed as well, but the binding of magnetite with chitosan was not strong enough in protonated chitosan solution, so chitosan hydrogel, instead of chitosan solution, has been used in the synthesis of Fe_3O_4 composite so far [40].

At first stage of our work the freshly prepared Fe_3O_4 particles were coved by one of the polymer, CS or MAC. The mild conditions for the formation of magnetite monopolymer shells were used: aqueous media, room temperature and ultrasonic treatment of reaction mixture. By preliminary experiments it was found that the pH optimum for IPEC formation was 3.5 for chitosan and for MAC – 5.5, and optimal concentration of solutions of polymer was 1-2%. As to poly (ethylene-*alt*-maleic acid), the concentration used did not exceed its critical concentration of aggregation (below 2% wt.) [41]. It was noted that more aggregated particles were formed at higher polymer concentrations and lower mixing rates. MNPs and MMPs fractions were separated by centrifugation. Table 1 shows the composites of magnetite covered by one of two polymers only and their characteristics (samples *1-6*). The contents of organic (polymer) phase in the composites magnetite/chitosan (Fe_3O_4/CS), magnetite/poly(ethylene-*alt*-maleic acid)(Fe_3O_4/EM), magnetite/poly(N-vinyl-2-pyrrolidone-*alt*-maleic acid) (Fe_3O_4/VM) were in the range of 24-43 % wt. and were mainly depended on sizes of particles but not on nature of polymer. MNPs contained maximal amounts of polymer (39-43% wt.) in shells since cores of MNPs were smaller in sizes and greater in the surface area than those of MMPs.

Table 1. Composition and Properties of Magnetite Composites

Sample	Particle type	Particle mean size (standard deviation)	Polymer content: Inner layer/Outer layer (wt. %)[f]	Saturation magnetization, (gauss·cm^3/g)
1	Fe_3O_4/CS	25(13)[c]/48(7)[d] nm	39,0/-	12.5
2	Fe_3O_4/CS[a]	15(10)[e] μm	29.2/-	48.3
3	Fe_3O_4/VM	38(18)[c]/63(8)[d] nm	42.1/-	12.1
4	Fe_3O_4/VM	22(10)[e] μm	24.0/-	39.0
5	Fe_3O_4/EM	25(15)[c]/54(9)[d] nm	43.2/-	12.0
6	Fe_3O_4/EM[a]	16(13)[e] μm	29.5/-	43.0
7	Fe_3O_4/CS/VM	34(14)[c]/75(10)[d] nm	38.0/33.0	6.5
8	Fe_3O_4/CS/VM[b]	28(11)[e] μm	19.3/1.2	49.2
9	Fe_3O_4/CS/EM[a]	26(12)[e] μm	18.0/2.7	-
10	Fe_3O_4/CS/EM	25(15)[e] μm	23.7/2.1	41.0
11	Fe_3O_4/CS/EM	88(10)[d] nm	39.0/28.0	-
12	Fe_3O_4/CS/EM[a,b]	22(14)[e] μm	20.4/1.3	-
13	Fe_3O_4/VM/CS	20(11)[c]/70(9)[d] nm	41.7/29.2	-
14	Fe_3O_4/EM/CS	23(12)[c] nm	40.0/30.0	-

[a] Fluorescently labeled sample. [b]Cross-linked composite shell. [c]AFM method. [d]DLS method. [e]Optical microscopy method (the samples also contained the particles less than 5 μm in diameter). [f]The composition was determined by elemental Fe, C, H, N analysis.

It should be noted that *in situ* technique used for the preparation of magnetite/CS (synthesis of Fe_3O_4 particles in the presence of polymer) resulted in higher heterogeneity in particle size distribution of the composites and lower magnetic saturation values in comparison with the method described above. Moreover *in situ* technique of stabilization of magnetite by MAC resulted in the precipitation of ferric-MAC salts. As a result we refused *in situ* technique in our work.

In contrast to monopolymer covering of Fe_3O_4 the bilayer one could be diminish the susceptibility of composites to the external factors (pH, ionic strength, etc.). The creation of interpolyelectrolyte complex shells is one of the ways of stabilization of different nano- and microparticles. There are several procedures for the preparation of IPEC: polymerization of ionic monomers on oppositely charged polymeric matrices or mixing of aqueous solutions of polycations and polyanions.

Earlier, for stabilization of magnetite an interpolyelectrolyte complex of CS with monocarboxylic acid polymer (polyacrylic acid – PAA) was used [33, 34]. Magnetic Fe_3O_4/CS/PAA microspheres were prepared by graft-polymerization of acrylic acid onto Fe_3O_4/CS particles. CS–PAA complex proved to be a good stabilizer of the well-dispersed suspension of paramagnetic Fe_3O_4.

In comparison with the described method [33], in the present work the outer (second) layer of interpolyelectrolyte magnetite shell was formed in water solution by means of adsorption of the oppositely charged polymer onto magnetite/monopolymer composite particles. Particles of *Type 1* contain MAC in the outer layer of interpolyelectrolyte shell, while particles of *Type 2* contain MAC in the inner one (Figure 1, Table 1, samples *7-14*). The concentrations of polymers used in the construction of both inner and outer layers of interpolyelectrolyte shell were identical, and pH values used for the formation of outer layer were in the range of 4-5. As it was previously shown, pH-optimum for formation of polyelectrolyte complex CS–MAC was about 4-5 since the degrees of ionization of both interacting amino and carboxylic groups under these conditions are high. At pH=5 the degree of protonation of amino groups of chitosan is 98%, and the degrees of ionization of first and second carboxylic groups of EM are 97% and 5%, respectively. The same values for VM are 95% and 2%, respectively [42]. Also it has been found that the process of CS and MAC interaction is cooperative, and the effective binding constants for reactions of CS with poly(N-vinyl-2-pyrrolidone-*alt*-maleic acid), poly(styrene-*alt*-maleic acid)or poly(ethylene-*alt*-maleic acid) are in the range of $(1 \div 3) \times 10^{-4}$ L/mol [43].

Prepared composites contained 25-75% of magnetite and exhibited the saturation magnetization values of 6-50 gauss$\times$cm^3/g. The latter values decreased with the increasing of the content of organic phase in the composites (Table 1). Magnetization values were low for Fe_3O_4 MNPs coved by one of polymers (samples *1, 3, 5*) only and they were minimal for MNPs stabilized with interpolyelectrolyte shells (sample *7*). The saturation magnetization values of MMPs containing large magnetic core were maximal (samples *2, 4, 6, 8, 10*)

Thermal Activation and Crosslinking of Magnetite Shells, HRP Immobilization on Activated Composite Particles

For stronger fixation of interpolyelectrolyte coating of composite particles their thermal treatment was carried out. The heating causes the transformation of ionic bonds between amino groups of CS and carboxylic groups of MAC to covalent amide (imide) bonds. Under these conditions dicarboxylic acid residues of outer copolymer layer untouched during the interpolyelectropolyte interaction were converted to anhydride groups [44] (Figure 1; FT-IR spectra (Figures 2, 3)). Another method for preparation of cross-linked, anhydride containing shell was the formation of outer layer on Fe_3O_4/CS composite particles in the waterless organic solvent (DMF) with the use of MAC taken in the form of maleic anhydride copolymer. The proposed method is a simple alternative approach for increasing of stability of shells by means of their covalent cross-linking with bifunctional agents or by using of condensing agents [34, 45].

The possibility of a practical application of the composites for biotechnological purposes was demonstrated for MMPs containing the interpolyelectrolyte shell. We prefered MMPs instead of MNPs because of more convenient working conditions in the case of using of MMPs - faster sedimentation of MMPs in the magnetic field and easier separation of liquid phase. Functional groups exposed on the outer surface of MMPs – Fe_3O_4/CS/VM (Table 1, *8*) and Fe_3O_4/CS/EM (Table 1, *11*), were activated by heating before enzyme immobilization. The reactivity of resulting anhydride groups of MMPs allowed the binding of horseradish peroxidase (HRP) via amide bonds formation under mild conditions in aqueous medium without the usage of any condensing agents. The contents of HRP immobilized onto Fe_3O_4/CS/EM and Fe_3O_4/CS/VM were found to be 10mg/g (0.25 μmol/g) and 1.6 mg/g, correspondingly; the relative specific HRP activity was 80% and 75%, respectively. Earlier described magnetized silica-based microparticles (5 μm) contained immobilized HRP in amount of 0.1-0.3 μmol/g [46].

Characterization of Magnetite Composites

Figure 2 (A) shows FT-IR spectra of magnetite coved by one of the polymers, IPEC shell and cross-linked shell. The resolutions of FT-IR spectra as regards to characteristic absorption bands of polymers in composites are not very clear since composite particles contain large amounts of Fe_3O_4. To clarify FT-IR spectra of organic phase of composites the model samples, containing only the polymers of shells, were prepared and analyzed. Figure 2 (B) shows FT-IR spectra of the model compounds – stoichiometrical IPEC, heated IPEC and copolymer of maleic anhydride.

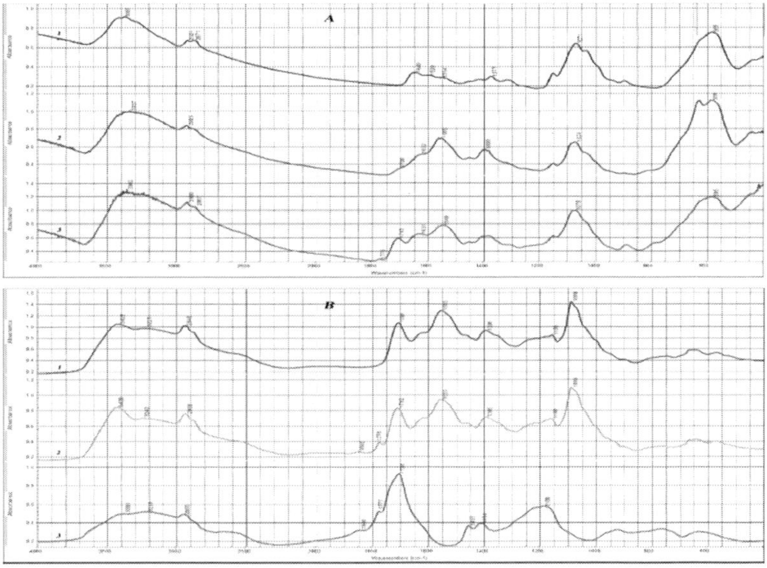

Figure 2. (A). FT-IR spectra of magnetite composites: Fe_3O_4/CS (*1*), Fe_3O_4/CS/EM (*2*), Fe_3O_4/CS/EM after thermal treatment (*3*). (B) FT-IR spectra of model compounds IPEC CS/EM (*1*), IPEC CS/EM after thermal treatment (*2*), EM (in the form of copolymer of maleic anhydride) (*3*).

The peaks at ν = 585-588 cm^{-1} in all spectra (Figure 2 (A)) were assigned to characteristic band of Fe–O within Fe_3O_4 [47]. The spectrum of Fe_3O_4/CS (Figure 2 (A), *1*) contains the characteristic absorption bands at 3400-3200 cm^{-1} which were assigned to the stretching vibrations of OH and NH. CH stretch vibrations were at 2921, 2871 cm^{-1}, C–O–C – at 1071 cm^{-1}, 1649 cm^{-1} (amide I), 1554, 1599 cm^{-1} (amide II) and 1377 cm^{-1} (CH_3) (CH_3CONH-groups of chitosan). IR spectrum of CS-coated magnetite particles was described earlier [48]. In the spectrum of Fe_3O_4/CS/EM (Figure 2 (A),*2*) and IPEC CS/EM (Figure 2 (B), *1*) the absorption bands in the region of 1632-1708 cm^{-1}, ν_{COO^-} =1555, 1400 cm^{-1}, and ν_{C-O}=1074 cm^{-1} belonging to CS–MAC interpolyelectrolyte complex were presented earlier[44]. The thermal treatment of Fe_3O_4/CS/EM (Figure 2 (A), *3*) and CS/EM (Figure 2 (B), *2*) led to cross-linking of shell CS/EM. As a result the absorption band of amide groups at 1549 cm^{-1}, the adsorption band at $\nu_{C=O}$=1710 cm^{-1} and 1776 cm^{-1} appeared. The last two bands indicate the appearance of the cyclic structure of succinic anhydride and partially succinimide groups similar to those which appeared after the thermal treatment of CS–MAC interpolyelectrolyte complex [44]. The cyclic structure of succinic anhydride was sharply defined also in the spectrum of poly (ethylene-*alt*-maleic anhydride) (Figure 2 (B), *3*).

Transmission electron microscopy (TEM) shows the presence of magnetite nanocrystals and nanoparticles in the form of flocculated clusters in cores of composite particles after all steps of formation of polymer shells. Histograms of size distributions of Fe_3O_4 nanocrystals (individual ones and in clusters) in the mono- and bilayer covered magnetite particles are represented in TEM images (Figure 3, *a1, b, c*). Nanocristals of magnetite in all magnetite-polymer composites are roughly spherical and their sizes range in 3 nm $< D_{TEM} <$ 15 nm. The formation of iron oxide clusters of individual nanocrystals is governed by a balance of van der Waals and electrostatic interactions. Moreover, polymers are able the clustering of the primary magnetite nanocrystals upon their interaction with polymer chains. Also, it is necessary to take into consideration the opportunity of alteration of cluster sizes during the preparation of samples for TEM, since their morphological appearance can be perturbed from that in the aqueous dispersion. The sizes of discovered clusters in cores of the samples varied from 20 to 400 nm (Figure 3, *a2*) independently of stage of formation of polymer shell.

The cores of MNPs apparently contained individual nanocrystals and small nanoclusters while MMPs ones contained the clusters of Fe_3O_4 mainly. The differences in ratio of the amounts of nanocrystals and clusters were in accordance with these (about 1: (8-9)) of amounts of MNPs and MMPs obtained at the experimental conditions used (polymer concentrations, pH of medium, velocity and time of mixing).

The crystalline nature of nanoparticles of Fe_3O_4, for example, for composite particles Fe_3O_4/CS/VM is illustrated by the electron diffraction (ED) image (Figure 3, *c*, insert). The strong lattice rings in ED image can be indexed to (220), (311), (400), (422), (511) and (440) planes of the standard magnetite system (JCPDS card No: 19-0629).

As it was expected, average size (diameter) of magnetite composite particles reasonably exceeded that of metal oxide particles themselves. Thus, the average size of MMPs (optical microscopy data, see Figure 4) in the case of magnetite with monopolymer shell was 15-20 μm (standard deviation of 10 μm) (Table 1, *2, 4, 6*), and that of Fe_3O_4/CS/VM and Fe_3O_4/CS/EM 22-28 μm (standard deviation of 11-14 μm) (Table 1, *8-11*). As it is seen after application of second polymer layer on magnetite/monopolymer microparticles small increasing in size of particles was occured only.

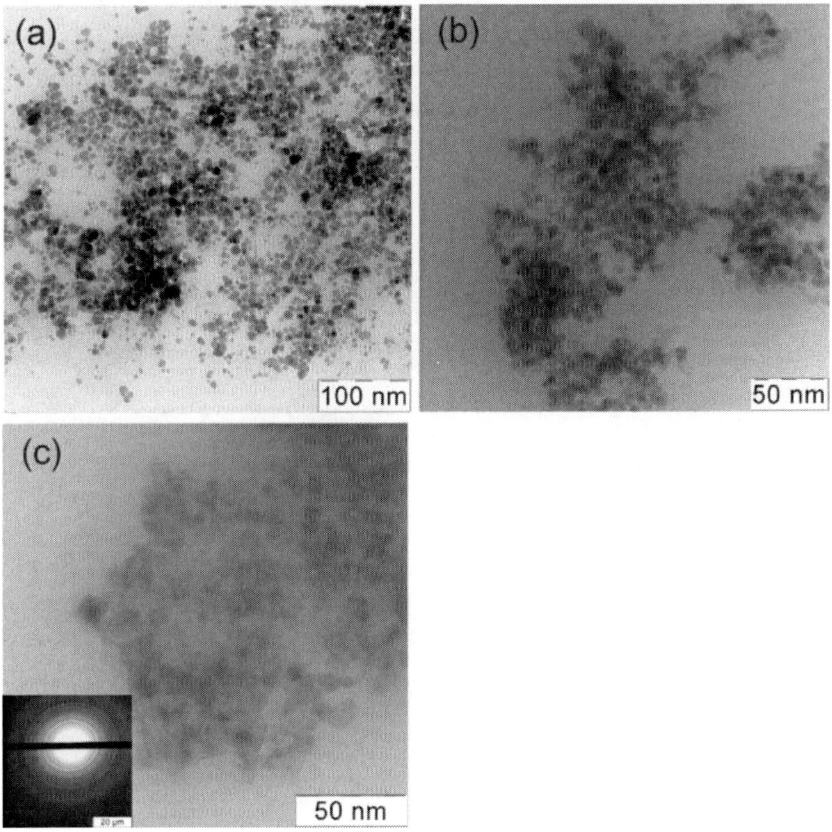

Figure 3. TEM micrographs of magnetite composites: (a1, a2) Fe_3O_4/CS (Table 1, *1*), (b) Fe_3O_4/VM (Table 1, *3*), (c) Fe_3O_4/CS/VM (Table 1, *7*), inset shows the electron diffraction pattern. Size distributions: nanocrystals (a1, b, c) and clusters (a2).

Table 2. Content of Polymer in Outer Layer of Magnetite Composite Interpolyelectrolyte Shell by the Dye Displacement Method

Initial composite (sample number, Table 1)	Polymer content in initial composite, mmol/g	Polymer of outer layer	Polymer content in outer composite layer by the dye displacement method, mmol/g
Fe_3O_4/CS (*1*)	2.21	VM	1.75
Fe_3O_4/CS (*2*)	1.65	VM	0.041
Fe_3O_4/VM (*3*)	2.38	CS	1.70
Fe_3O_4/VM (*4*)	1.05	CS	0.034
Fe_3O_4/EM (*5*)	2.98	CS	1.80

The sizes of MNPs determined by DLS are represented in Table 1 (samples *1, 3, 5, 12*). Sizes of double-layered MNPs also slightly exceed the sizes of mono-coated nanoparticles. The data of DLS sizes determined for MNPs naturally exceeded the size of these particles determined by AFM method.

Figure 5 shows AFM images of magnetite composites containing both monopolymer and interpolyelectrolyte shells. It is seen from the images that the shapes of most of the particles were almost spherical and the average diameters of particles were in the range of 5-100 nm. The whole array of particles is divided into chosen size ranges (Table 3). Table 3 shows that the sizes of magnetite particles stabilized by one of polymer alter slightly after the treatment of the particles with an oppositely charged polymer, as it was shown by DLS studies and optical microscopy.

Apparently the second layer was formed by the polymer molecules in their expanded conformations, and this layer did not form "braches" or "mushrooms" on the surface of particle. Here we could assume that during the formation of IPEC amino groups of chitosan glucosamine units and carboxylic groups of maleic acid residues of MAC are not quantitatively involved into interpolyelectrolyte interaction. This assumption is supported by the model of stoichiometrical IPEC complex (Figure 1), which takes into account the regularity of the structure of macromolecular chains of the interacting polymers and their steric conformation. Also this model takes into considerations the distances between NH_2-groups of two neighboring units of glucosamine in chitosan (≈5.15 Å) [49] and the distances between maleic acid residues of MAC (≈5.02 Å). The latter value is calculated for the "ideal" polymer chain, where the distance between the chain carbon atoms is 1.54 Å and the angle between the links 109°, assuming that the polymer chain is similar to that of polyethylene. In accordance with the model, only every second amino group of chitosan corresponds sterically and can be bound with the carboxyl group of the every second residue of maleic acid in MAC only.

Real molecular structure of polymers in the solution is sensitive to solvent, temperature, ionic strength, and pH. For example, the dichroism spectrum of chitosan-aspartic acid complex indicates that chitosan adopts the helical (left-handed) conformation [50]. N.Ohno described the pH-induced conformational transitions of maleic acid and α-methylstyrene copolymer from the compact form to the coil [51].

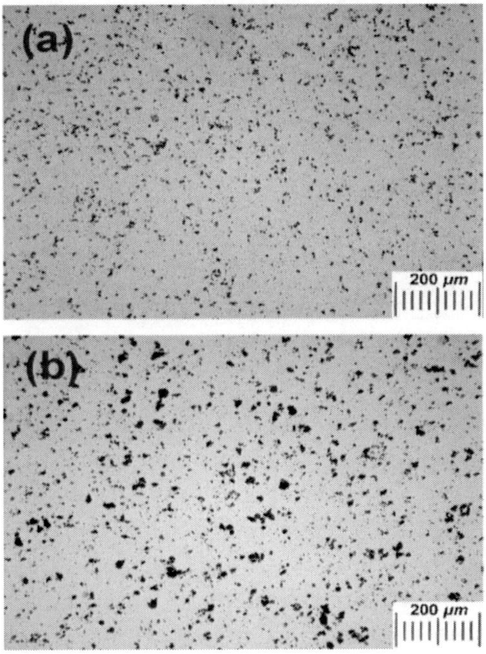

Figure 4. Optical microscopy micrographs of magnetite nanocomposites: (a) Fe_3O_4/CS (Table 1, *2*), (b) Fe_3O_4/CS/EM (Table 1, *10*).

For a number of magnetite/monopolymer composites the effective amount of second polymer layer which can be involved in interpolyelectrolyte shell is assessed by the dye displacement method. Amino groups of CS of composite Fe_3O_4/CS displaced the methylene blue (MB) from VM–MB salt into solution, and carboxylic acid groups of VM of Fe_3O_4/VM composite displace eosin (E) from CS–E salt. The maximal optical density of the dye was used as a point of saturation of Fe_3O_4/monopolymer composite shell by the second polymer. And at this point the effective amount of second polymer layer was applied onto the surface of magnetite/monopolymer composite particles.

Table 3. Particle Size Distribution from the AFM Data for MNPs with "Monopolymer" and Interpolyelectrolyte Coatings

Composite (sample number in Table1)	Size range, nm	Mean radius (standard deviation), nm	Particle content, %
Fe_3O_4/EM [5]	5-20 20-100 100-200	10 (5) 43 (19) 120(13)	61 36 2
Fe_3O_4/EM /CS [14]	5-20 20-100 100-200	11 (5) 36 (17) 107 (5)	55 43 2
Fe_3O_4/VM [3]	5-20 20-100 100-200 >200	17(2) 38(18) 141(28) 316(155)	8 72 12 8
Fe_3O_4/VM /CS [13]	5-20 20-100 100-200	9(4) 31(11) -	75 25 -
Fe_3O_4/CS [1]	5-20 20-100 100-200 >200	12(5) 39(18) 134(33) 531(230)	55 43 2 1
Fe_3O_4/CS/VM [7]	5-20 20-100 100-200	10(5) 39(14) 131(31)	67 31 2

In these experiments we use diluted solutions of polymers. Surface of magnetite/monopolymer composite particles after the ousting of the dye bound the oppositely charged polymer. The results show that the amount of introduced second polymer layer increased with decreasing of particle size from micro - (Table 2, samples *2, 4*) to nanoparticles (Table 2, samples *1, 3, 5*). In some cases a part of the total amount of the first polymer layer is able to participate in the interpolyelectrolyte interaction only. In the case of MMPs (samples *2, 4*), about 2.5–3.0 % of total amount of polymer of the first layer are connected with the second polymer layer. This approach allows assessing the amount of oppositely charged polymer required for the construction of outer polymer layer taking into account the size of magnetite/monopolymer composite particles.

To evaluate the amount of functional groups which are not involved into interpolyelectrolyte interaction and can be used for modification the titration of MMPs by the anionic (eosin) or cationic (methylene blue) dyes was carried out. The titration showed that a sufficient amount of functional groups of outer polymer layer available for connection with the dye. It should be noted that some amount of the functional groups of inner layer of magnetite interpolyelectrolyte shell also were available for titration. For example, even in the case of cross-linked MMPs Fe_3O_4/CS/EM (after hydrolysis of anhydride groups) 12.4 μmol/g of maleic acid residues and 0.8 μmol/g amino groups were revealed. Previously it was shown that during the covering of different surfaces with the polyelectrolyte complex composed of chitosan and poly (N-vinyl-2-pyrrolidone-*alt*-maleic acid) by layer-by-layer method, it was shown that the outer copolymer layer contained sufficient amount of carboxylic groups, capable of bearing the specific to plasminogen ligand [44].

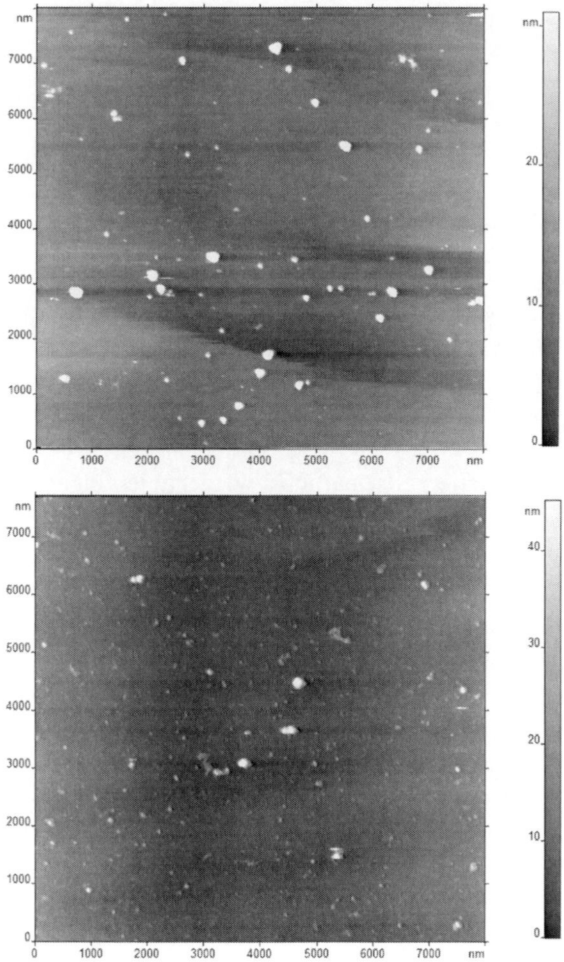

Figure 5. AFM images of magnetite nanocomposites: Fe_3O_4/EM (Table 1, *5*) – left image; Fe_3O_4/EM/CS (Table 1, *14*) – right image.

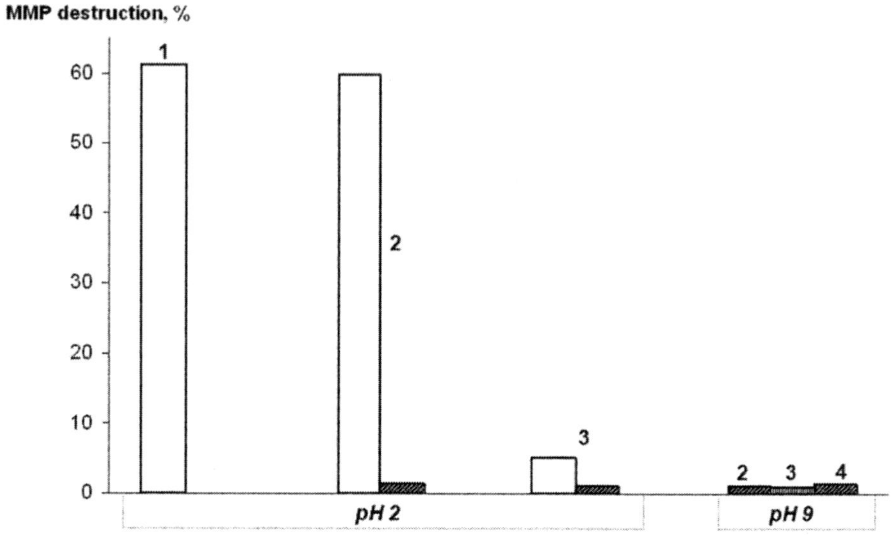

Figure 6. Magnetite microparticles shell destruction (100 hours): 1 – Fe_3O_4/CS (Table 1, *2*), 2 – Fe_3O_4/CS/EM (Table 1, *9*), 3 – Fe_3O_4/CS/EM (cross-linked) (Table 1, *12*), 4 – Fe_3O_4/EM (Table 1, *6*). Histogram bars: white columns – CS–SR detection, black columns – FMOC–EM detection.

The obtained magnetite microparticles were stable longer than 6 months at physiological values of pH, and MMPs were more stable then MNPs. Presence of negative or positive charges on nanoparticles were beneficial for their colloidal stability because of electrostatic repulsion forces. Previously it was shown earlier that maghemite NPs protectively coated with IPEC in two layer-by-layer deposition steps with poly(ethylene imine) and poly[(ethylene oxide)-block-poly(glutamic acid)] were stable in physiological salt solution for longer than 6 months [52].

The stability of prepared in our work MMPs covered by one of the polymer or interpolyelectrolyte shell was also evaluated under conditions, more severe then these at physiological values of pH. At first, stability of MMPs has been tested in highly acidic medium in which both parent polymers are soluble and the degradation of the IPEC is occurred [42]. Additionally the measurements were carried out in highly alkaline aqueous solution, where MACs are soluble while chitosan is insoluble. For these experiments the parent polymers were preliminary labeled and then they were applied for covering of magnetite. The stability was defined as a percentage of content of the labeled polymer in solution to its content in the initial MMPs. Histogram

(Figure 6, *1*) shows that a significant degradation of polymer shell of Fe_3O_4/CS took place during 100 h at pH 2 (labeled CS was monitored in the solution). In the case of composite Fe_3O_4/CS/EM (Figure 6, *2*, pH 2) the degradation of interpolyelectrolyte coating was somewhat lower. As it was shown above (the composite titration experiments) both monopolymer and interpolyelectrolyte magnetite coatings contained unbound amino groups of CS, and these groups apparently contributed to the hydrolytic instability of composites. The cross-linked composite Fe_3O_4/CS/EM (Figure 6, *3*, pH 2) was found to be practically stable at these conditions.

The same tendency (but significantly less pronounced) was observed during the hydrolysis of composites when the labeled copolymer was monitored in the solution (Figure 6, black columns). Low values of degree of degradation estimated by monitoring of labeled copolymer in the solution can be explained by a strong adhesion of carboxylic groups of polymer to magnetite [39, 53]. It was shown that carboxylic groups of polyacids form relatively stable complexes with atoms of iron of magnetite through bidentate bonding [54]. Besides the destruction of protective coating may cause the oxidation of magnetite as well:

$$2\ Fe_3O_4 + 0.5\ O_2 + 9\ R(COOH)_2 = 3\ (Fe^{+3})_2[R(COO^-)_2]_3 + 9\ H_2O,$$

and the liberated ions of iron can form insoluble complexes with residues of dicarboxylic acid of copolymer $R(COOH)_2$.

In alkaline media, CS was not detected in the solution. As to copolymer, its amount in the solution did not exceed 5% for all samples during the observation time (Figure 6, columns *2*, *3*, *4*, pH 9).

CONCLUSION

Here we reported a simple procedure for preparation of stabilized and functionalized magnetite nano- and microparticles by successive application of oppositely charged polymers having regular structures of macromolecular chains (chitosan and maleic acid copolymers) onto Fe_3O_4 core. The proposed approach makes it possible to create magnetite interpolyelectrolyte shells of two reversed types containing different functional groups on outer shell layer – carboxylic or amino groups. Also the effect of different factors on the formation, structure and properties of obtained magnetite particles coved by one of the polymers and interpolyelectrolyte shells is demonstrated. The structure and composition of IPEC shells, magnetic properties of composites depended on nature of maleic acid copolymer, reaction conditions and size of core of synthesized composite particles. The prepared composites contained 25-75% magnetite and exhibited the saturation magnetization values of 6-50 gauss ×·cm^3/g, and the latter value decreased with the increase of the content of organic phase in the composites (from micro- to nanoparticles).

The used analytical procedures allowed us to assess the amount of functional groups on the surface of composite particles. It was shown that the outer copolymer layer contained sufficient amount of functional groups available for subsequent modification. Surface-exposed maleic acid residues of copolymer were transformed to reactive anhydride groups by means of thermal activation followed by covalent cross-linking of the shell. Ion-stitched IPEC shells of magnetite microparticles were stable in aqueous media at physiological values of pH and the covalently cross-linked shells – in highly acidic and alkaline solutions.

Horseradish peroxidase was successfully immobilized onto activated and covalently cross-linked Fe_3O_4/CS/MAC microparticles in aqueous medium without usage of any outer condensing agents. The proposed reproducible and low-cost technique doesn't use toxic reagents and solvents at all stages, including the preparation of Fe_3O_4 and formation, cross-linking, activation of interpolyelectrolyte complex shells and modification of surface of composite particles. The method offers the opportunity for immobilization of different ligands or bioactive substances, containing amino or hydroxyl groups. Prepared MMPs are nonporous, and this their property minimizes the diffusion factors. High magnetic saturation values of MMPs make them easy to isolate and concentrate with the use of a laboratory permanent magnet without application of laborious separation methods such as chromatography, filtration and centrifugation. We believe that the techniques described here can be applied for stabilization and functionalization of nanoparticles of metals or their alloys.

ACKNOWLEDGMENTS

We are grateful to Z.S. Klemenkova for assistance with IR-spektra, Dr. S.S. Abramchuk for assistance with TEM, V.I. Lozinskii for help with light microscopy (all from Nesmeyanov Institute of Organoelement Compounds, Russian Academy of Sciences), Dr. V.A. Shandincev for technical support and investigation of magnetic properties of magnetic composites and A. S. Erofeev (Faculty of Physics, Moscow State University) for help with AFM.

REFERENCES

[1] Dave, SR; Gao, X. *WIREs Nanomed. Nanobiotechnol.*, 2009, 1, 583-609.
[2] Huang, S-H; Juang, R-S. *J. Nanoparticle Res.*, 2011, 13, 4411-4430.
[3] Giouroudi, I; Kosel, J. *Recent Pat. Nanotechnol.*, 2010, 4, 111-118.
[4] Wang, M; Thanou, M. *Pharmacol. Res.*, 2010, 62, 90-99.
[5] Peng, S; Cheng, K; Sun, S. *Chem. Soc. Rev.*, 2009, 38, 2532-2542.
[6] McBain, SC; Yiu, HHP; Dobson, J. *Int. J. Nanomed.*, 2008, 3, 169-180.
[7] Gervald, AY; Gritskova, IA; Prokopov, NI. *Russ. Chem. Rev.*, 2010, *79*, 219-229.
[8] Safarik, I; Safarikova, M. *Monatshefte fur Chemie.*, 2002, 133, 737-759.
[9] Indira, T. K; Lakshmi, P. K. *Int. J. Pharm. Sci. Nanotech.*, 2010, 3, 1035-1042.
[10] Babinkova, M; Babinec, P. *Biomed. Papers.*, 2009, 153, 243-250.
[11] Hsing, I-M; Xu, Y; Zhao, W. *Electroanalysis.*, 2007, 19, 755-768.
[12] Laurent, S; Forge, D; Port, M; Roch, A; Robic, C; Vander, EL; Muller, RN. *Chem. Rev.*, 2008, 108, 2064-2110.
[13] Srajer, G; Lewis, LH; Bader, SP; Epstein, AJ; Fadley, CS; Fullerton, EE; Hoffmann, AJ; Kortright, B; Krishnan KM, Majetich SA, Rahman, TS; Ross, CA; Salamon, MB; Schuller, IK; Schulthess, TC; Sun, JZ. *J. Magn. Magn. Mater.*, 2006, **307,** 1-31.
[14] Odenbach S (ed). Ferrofluids: Magnetically Controllable Fluids and Their Applications, (LNP 594). Springer-Verlag: Berlin- Heidelberg, 2002**.**
[15] Black, CT; Murrey, CB; Sandstom, RL; Sun, S. *Science.*, 2000, 290, 1131-1134.
[16] Zeng, H; Li, J; Liu, J. P; Wang, ZL; Sun, S. H. *Nature.*, 2002, 420, 395-398.
[17] Tiwari, S; Choudhary, RJ; Phase, DM. *Thin Solid Films.*, 2009, 517, 3253-3256.
[18] Escobar, GP; Beroy, AQ; Iritia, MPP; Huera, JH. *Chem. Eng. J.*, 2004, *102,* 107-117.
[19] Petrova, TM; Fachikov, L; Hristov, J. *Int. Rev. Chem. Eng.*, 2011, 3, 134-152.

[20] Tseng, J-Y; Chang, C-Y; Chen, Y-H; Chang, C-F; Chiang, P-C. *Colloid Surfaces A: Physicochem. Eng. Aspects*., 2007, 295, 209-216.
[21] Ngomsik, A-F; Bee, A; Graye, M; Cote, G; Cabuil, V. *C. R. Chimie*., 2005, *8*, 963-970.
[22] Martina, MS; Fortin, JP; Menager, C., Clement, O; Barret, G. Grabielle-Madelmont, C; Gazeau, F; Cabuil, V; Lesreur, S. *J.A.C.S*., 2005, 127, 10676-10685.
[23] Ito. A; Takizawa, Y; Honda, H; Hata, K; Kagami, H; Ueda, M; Kobayashi, T. *Tissue Eng*., 2004, 10, 833-840.
[24] Parak, WJ; Gerion, D; Pellegrino, T; Zanchet, D; Micheel, C; Williams, S. C; Boudreau, R; Le Gros, M. A; Larabell, C. A; Alivisatos, A. P. *Nanotech*., 2003, 14, R15-R27.
[25] Makhluf, SB-D; Abu-Mukh, R; Rubinstein, S; Breitbart, H; Gedanken, A. *Small*., 2008, *4*, 1453-1458.
[26] Shaw, S-Y; Chen, Y-J. X; Ou, J-J; Ho, L. *Enzyme Microbiol. Technol*., 2006, 39, 1089-1095.
[27] Schultz, N; Syldatk, C; Franzreb, M; Hobley, TJ. *J. Biotechnol*., 2007, 132, 202-208.
[28] Fuentes, M; Mateo, C; Rodriguez, A; Casquero, M; Tercero, JC; Reise, HH; Fernandez-Lafuente, R; Guisan, JM. *Biosensors Bioelectronics*., 2006, 21, 1574-1580.
[29] Ivanova, V; Petrova, P; Hristov, J. *Chem. Eng*., 2011, 3, 289-299.
[30] Balmayor, ER; Pashkuleva, I; Frias, AM; Azevedo, HS; Reis, RL. *J. R. Soc. Interface*., 2011, 8, 896-908.
[31] Lang, C; Schüler, D. *J. Phys.: Condens. Mater*., 2006, 18, 2815-2828.
[32] Samoilova, NA; Krayukhina, MA; Yamskov, IA. Interpolyelectrolyte Complexes of Chitosan. Nova: New York, 2010.
[33] Wu, Y; Guo, J; Yang, W; Wang, C; Fu, S. *Polymer*., 20**06**, 47, 5287-5294.
[34] Zhu, A; Luo, X; Dai, S. *J. Mater. Res*., 2009, 24, 2307-2315.
[35] Conix, A; Smets, G. *J. Polym. Sci.* 1995, 15, 221-229.
[36] Elmore, WC. *Phys. Rev.* 1938, 54, 309-310.
[37] Ge, Y; Zhang, Y; He, S; Nie, F; Teng, G; Gu, N. *Nanoscale Res. Lett*., 2009, 4, 287-295.
[38] Hasinoff, BB; Dunford, HB. *Biochemistry*. 1970, *9*, 4930-4939.
[39] Ge, J; Hu, Y; Blasini, M; Beyermann, WP; Yin. Y. *Angew. Chem. Int. Ed*., 2007, 46, 4342-4345.
[40] Wang, Y; Li, B; Zhou, Y; Jia, D. *Polym. Adv. Technol*., 2008, 19, 1256-1261.

[41] Samoilova, N; Kurskaya, E; Krayukhina, M; Askadsky, A; Yamskov, I. *J. Phys. Chem. B.*, 2009, 113, 3395-3403.

[42] Samoilova, NA; Krayukhina, MA; Yamskov, IA. *Russ. J. Phys. Chem. A.*, 2002, 76, 1502-1507.

[43] Krayukhina, MA; Samoilova, NA; Erofeev, AS; Yamskov, IA. *Polym Sci A.*, 2010, 52, 243-250.

[44] Samoilova, NA; Krayukhina, MA; Novikova, SP; Babushkina, TA; Volkov, IO; Komarova, LI; Moukhametova, LI; Aisina, RB; Obraztsova, EA; Yaminsky, IV; Yamskov, IA. *J. Biomed. Mater. Res.*, 2007, 82A, 589-598.

[45] Lai, GS; Zhang, HL; Han, DY. *Sens Actuat B.*, 2008, 129, 497-503.

[46] Yu, D; Blankert, B; Bodoki, E; Bollo, S; Vire, J-C; Sandulescu, R; Nomura, A; Kauffmann, J-M. *Sensors & Actuators, B.*, 2006, 113, 749-754.

[47] Li, G-Y; Huang, K-L; Jiang, Y-R; Ding, P; Yang, D-L. *Biochem. Eng. J.*, 2008, 40, 408-414.

[48] Nasra, MK; Mohamed, MM; Elblbesy, MA; Hefney, BA. *J. Biomat. Nanobiotech.*, 2011, 2, 193-199.

[49] Denuziere, A; Ferrier, D; Domard, A. *Carbohydr. Polym.* 1996, 29, 317-323.

[50] Singh, J; Dutta, PK. *Intern. J. Biol. Macromol.*, 2009, 45, 384-392.

[51] Ohno, N. *Polym. J.* 1981, 13, 719-726.

[52] Thünemann, AF; Schütt, D; Kaufner, L; Pison, U; Möhwald, H. *Langmuir.*, 2006, 22, 2351-2357.

[53] Cornell, RM; Schwertmann, U. The Iron Oxides: Structure, Properties, Reactions, Occurrence and Uses, Wiley-VCH: Weinheim, 2003.

[54] Lin, C-L; Lee, C-F; Chiu, W-Y. *J. Colloid. Interface. Sci.*, 2005, 291, 411-420.

INDEX

B

C

G

H

I

J

K

L

M

N

O

P

Q

R

S

T